KT-562-598

The River of Time

Igor D. Novikov

Translated from the Russian by Vitaly Kisin

CAMBRIDGE
UNIVERSITY PRESS

PUBLISHED BY THE PRESS SYNDICATE OF THE UNIVERSITY OF CAMBRIDGE
The Pitt Building, Trumpington Street, Cambridge CB2 1RP, United Kingdom

CAMBRIDGE UNIVERSITY PRESS
The Edinburgh Building, Cambridge CB2 2RU, United Kingdom
40 West 20th Street, New York, NY 10011-4211, USA
10 Stamford Road, Oakleigh, Melbourne 3166, Australia

© Cambridge University Press 1998

This book is in copyright. Subject to statutory exception
and to the provisions of relevant collective licensing agreements,
no reproduction of any part may take place without
the written permission of Cambridge University Press.

First published 1998

Printed in the United Kingdom at the University Press, Cambridge

Illustrated by K. Moshkin
Additional illustrations for Chapter 15
of the English edition by the author

A catalogue record for this book is available from the British Library

Library of Congress Cataloguing in Publication Data

Novikov, I. D. (Igor ' Dmitrievich)
The river of time / Igor ' D. Novikov.
p. cm.
Includes indexes.
ISBN 0 521 46177 4 - ISBN 0 521 46737 3 (pbk.)
1. Time. 2. Time-History. 3. Time-Philosophy. I. Title.
QB209.N68 1998
530.11–dc21 97-43010 CIP

ISBN 0 521 46177 4 hardback
ISBN 0 521 46737 3 paperback

To my children,
Elena and Dmitri,
who have longer than myself
to flow down
the river of time

I tell myself that like water
time flows between one's fingers
onto the sand that slowly cools,
and through the sand it seeps into nowhere...

    and if Styx is indeed a river
    that separates two worlds so far apart
    then its flow is lost among millennia.

Still, we know a river that has no bottom
one whose banks do not restrain its flow...
a moment comes when human names sink into it.

    Its waters are transparent and dark,
    they fill up everyone and everything,
    one can discern them between lines and hear them in music.

One wades into this river only once,
is banned from ever finding the mysterious source
where Time is fast asleep, curled in a tight cocoon
on the rocky bosom of Eternity.

Marina Katys

Contents

Preface to the Russian edition

The person to whom I owe my fate was my grandmother. My parents were not there to take part in bringing me up, so my first consciously made steps in life grew from her love and care. Once she found for me an exciting book: *Brer Rabbit's Adventures*, translated into Russian. I learnt to read with this book. It was my grandmother again who bought for me, on a flea-market, my first popular book about science. It was a very difficult time, the Second World War was raging and the family was evacuated to the town of Krasnokamsk on the Volga. People thought about food first, books were very secondary. But my grandmother – mind you, she had no education whatsoever – felt, perhaps, that food for thought was just as necessary for kids as food for the stomach. The book that she bought (or swapped?) was marvelous; I will never forget it. It was *Children's Encyclopaedia*, a pre-1917 book, with wonderful color prints. As far as I can remember, their quality was far superior to the often smeared and bleak illustrations that I find nowadays in some editions of books that I write.

That book had a chapter about astronomy. Browsing for the first time through the volume (as for any other kid, this was the first thing to do with a new book), I was amazed by a drawing of a gigantic fountain of fire, with a small globe of our Earth alongside.

I learnt later that it was a solar protuberance, and that the Earth was placed there for comparison. The image was so grandiose that I was in absolute awe. I was impressed by the majestic scale of natural processes which were much grander than anything that my childish imagination could conjure up.

Truly, that print proved to be auspicious for me. It was enigmatic, baffling and mysteriously attractive. I very quickly read everything it contained about astronomy, and then all the other chapters. Some sections on world history were quite interesting, but nothing could compare with astronomy! The depths of cosmic space, the vortices on the Sun, and the possibility of life on Mars captured my inquisitiveness, my imagination and my love. I think that the mysterious phenomena of the Universe were the fount of all these feelings. I knew they were for life. 'The light of the first love is in each of us.'

Life can display so much, it is so multifaceted and wonderful, but it can also be terrible. I lived with my grandmother because my father, who occupied a responsible position in the People's Commissariat for Transportation, was arrested in 1937 and 'died in prison' (according to an official acknowledgement, that is; in 'their' parlance, this stands for 'was executed'), while my mother was deported to exile. Both were completely cleared of all accusations ('rehabilitated', in Soviet parlance) in the 1950s. Nevertheless, I did not know and still do not know of anything more wonderful than striving to learn the mysteries of the Universe. What I mean is not an abstract longing, not a lazy 'philosophizing' about the meaning of existence (I understood quite early that this was nonsense and, often, a manifestation of laziness and self-admiration caused by each wiggle of one's thought) but hard and happy work.

From early childhood I grew more and more certain that the best way to stimulate the development of the mind and of its creative potential is to strike a spark of unstoppable inquisitiveness into the mysteries that nature is hiding. True inquisitiveness will lead one further, make one seek and toil, even if he or she never becomes a scientist.

Later I read a great many science-popularizing books. Frankly, their number was much tinier then than now but... most of them were quite good! I learnt very soon that one needs to know an awful lot if one wishes to really accomplish anything in science. The fire of inquisitiveness was burning in me, so nothing could ever stop me. Furthermore, years of studying, of overcoming small but gradually more difficult obstacles, were rewarded with constantly growing delight.

Why am I telling all this?

I do it to illustrate two ideas with my own fate. Firstly, it is extremely important to imbue a person with a bona fide scientific thirst for knowledge, which later will become this person's driving force. It is not essential that he or she actually becomes a professional scientist. A love of science, a comprehension of its foundations, an admiration of the discoveries that unravel the most profound secrets of nature are as necessary to any person as an all-round cultural and aesthetic education. Our contemporaries cannot live without music, or paintings, or books. A life without appreciation of the achievements of science, which comes up with answers to the most profound whys and hows that we ask of nature, is equally unacceptable. A well-known physics theorist in the USSR, Vitaly Ginzburg has said this about the theory of relativity – one of the most perfect physical theories of our time: it incites 'a feeling... akin to what one feels looking at the most outstanding masterpieces in painting, sculpture or architecture'.

I will also cite a Soviet philosopher Boris Kuznetsov discussing the art and science of ancient Greece as the unifying elements of human culture: 'it speaks... of life uninterrupted, of new impressions, feelings and thoughts that are still inspired by Venus of Milo or Nike of Samothrace. In the same vein, we perceive the immortality of Plato's *Dialogues* or Aristotle's *Physics*.'

Secondly, to become a physicist or astronomer and really participate in scientific progress, one has to master the entire body of knowledge in the field one has chosen. Dilettantism has no place here. Science of today is incredibly complex and its mathematical equipment is so abstract and abstruse that the non-initiated simply could not fathom the degree of complexity of the whole. Actual work in science demands that you become an expert in applying the mathematical tools. Your knowledge of contemporary mathematics and related fields must be profound. This is the only level of expertise that allows one to reach the essence of subjects studied in physics and astronomy.

For a number of reasons, this level is not open to just anyone wishing to climb to it. Only a few become physicists, quite a few only handle mathematics within a high school course. Does it mean that any opportunity to admire the awesome achievements of physics is forever closed for these people, that it is impossible to find out about the science which penetrates the mystery of how matter is structured at its deepest levels and at the same time discovers the quanta of time and space?

Of course it does not, and one can describe the achievements of physics clearly and correctly to anyone interested, even without resorting to arithmetic. It means, however, that one should not try to explain all the details and difficulties in calculations and all the logical relations that lead to drawing the conclusions. The strategy must be different: one must try to create a shining image of a phenomenon, to make the reader form an idea of what the physicists attempt to achieve. These images can be understood without mathematics and can be admired and applauded. Remember, however, that if you are not a professional, not a physicist, do not entertain the illusion that having read a popular physics book you may be able to offer a 'hypothesis' that would solve the difficulties outlined in the book. Nothing good will come of it. An image is definitely not 'her majesty physics'. To offer a useful hypothesis, one has to become professional; however, everyone can enjoy an image drawn by a professional.

By way of comparison, I can say that I love music passionately but that God did not grant me a musical ear. I will never write music, nor reproduce even an elementary tune. I do enjoy listening to music written by (talented) professionals and performed by equally professional (also talented) individuals, and will continue to do so.

People who cannot draw or paint at all, do enjoy paintings, those who could not write a novel enjoy reading novels.

It is my firm belief that a similar situation holds for attempts to make science understood by the non-scientist. The author's goal must be to create a strong, impressive image.

I will try to describe in the subsequent chapters the achievements of physics that I dearly love.

This is a book about time, or rather, about scientists' attempts to understand what time is. The reader can be expected to ask, with full justification, whether there should exist a science of time. Isn't time something that anyone understands? What can one study about time?

I propose that you try to give a definition of what time is; I believe you won't be able to do it. Saint Augustine (354–430 AD) wrote: 'I know perfectly well what time is, as long as I do not think about it. But once I start thinking hard – I feel at a loss and do not know any more what time is.'†

Is it not true that anyone attempting to find an answer to this question feels a similar confusion? When we begin thinking about

† Translated from the version in Russian.

the nature of time, we tend to feel that this is an irresistible flow into which all events are embedded. Millennia of human experience seem to have proved that time flows at an unchanging pace. Apparently, it cannot be slowed down or accelerated. What is even more certain is that it cannot be turned back. For a long time, the notion of time remained a mere intuitive feeling and the object of abstract philosophical exercises.

In the first years of the 20th century it became clear that time *can* be influenced! For example, very fast motion slows down the pace of time. Next it was found that time flow is also affected by the gravitational field. An inseparable relation was discovered between time and the properties of space. This was the birth and the beginning of the rapid development of what we may call the physics of time (and space). Discoveries have been made recently in elementary particle physics and in astronomy, which greatly advanced our knowledge of the fascinating properties of time and may have brought closer the solution of the puzzles involved (for instance: why is a chain of events invariably one-dimensional but does not have, say, a 'width' and 'height' to which we are used in our three-dimensional space?; what was there before our Universe emerged? etc.).

The current stage in physics is characterized by a new and powerful breakthrough in our understanding of the structure of matter. In the first decades of the 20th century, physicists succeeded in unraveling the structure of atoms and in finding the main features of the interaction between atomic particles. Now physics studies quarks, which are subnuclear particles, and penetrates deeper and deeper into the microscopic world. All this progress is connected most closely with understanding the nature of time.

The book describes how the thinkers before us defined time and how the discoveries were made which showed that we may influence the flow of time. It describes how time flows in specific regions of the Universe, how it slows down in the neighborhood of neutron stars, how time is stopped in black holes and 'splashes over the brim' in white holes, how time may 'convert' into space and vice versa.

The properties of time are especially interesting at the first moments of the explosion which started the creation of our Universe; this was the period when time existed in the form of distinct time quanta.

The properties of time in superhigh-energy physics are important for science in general and for future technologies. Some very recent publications indicate that it might be possible to design a time machine which would allow time travel into the past.

The book also describes people who created the physics of time and who are doing further research in it now. It seems to happen too often that the great thinkers of the past or the distinguished contemporary scientists exist for the reading public only as abstract names mentioned in textbook and non-textbook publications, all written in a dry and very unemotional style. The images of these individuals are hardly associated with flesh-and-blood people, their interests, passions and contradictions. When I speak in this book about the scientific creativity of these scientists, I try also to find features and events that describe them as real human beings. On the other hand, it was never my intention to give their detailed biographies or to list their scientific achievements.

The book is aimed at readers interested in the history of scientific ideas, in the puzzles facing contemporary science, and in the personalities of scientists themselves, especially those physicists that I have had the pleasure and honor of meeting and working with. I do not assume that a reader has any special knowledge beyond the simplest course of high-school physics.

The reader will find that I chose a personalized style of presentation, especially when outlining studies in which I participated myself or when describing my meetings with physicists and astronomers. I quote in this connection Professor Vitaly Ginzburg, who had said this about one section of his scientific paper:

> It is not customary to use 'I' and 'me' in the scientific literature, especially in the Russian language. The same is mostly true for the science popularizing literature, so that the author has been referring to himself above only as 'we' or 'us' or was using other turns of speech suitable for such occasions. It would be difficult and even strange, however, to keep to this style in this specific section of the paper, since it is to a large extent autobiographical... I hope, therefore, that several personal pronouns will not produce intense negative response among the readers.

I hope that neither will my readers judge me too harshly for this 'immodest' presentation of my personal thoughts and my impressions.†

To write this book, I had to draw in places from my earlier popular physics and astrophysics books; some of them were written in collaboration with other people, to whom I express my gratitude.

The book cites a considerable number of quotations. Quite often, these are little known pronouncements of outstanding scientists of the past, as well as our contemporaries. I firmly believe that only the exact words of these illustrious personalities can bring to the reader their thoughts (and quite often their feelings as well). The great Russian poet Aleksander Pushkin said: 'To follow the reasoning of a great personality is a most captivating and gratifying subject' *(Arap Petra Velikogo* [Peter the Great's Moor]).

I. D. Novikov
Moscow

---

† *Remark for the English translation:* This self-justification may sound strange to an English reader. As far as I can see, authors of science-popularizing books have no qualms in using the pronouns 'I', 'me' or 'myself' when appropriate. This is not the case, however, for Russian literature.

Preface to the English edition

I began preparations to publish this book in English at the end of 1991; for a number of reasons, this stage stretched to several years. An oriental adage says: 'Hours tick away, days run away but years fly away.' These words are a reflection of our subjective perception of time intervals in the past, of what we remember of them. For most people, the feeling of the flight of time is considerably intensified when one turns in one's mind's eye to larger and larger blocks of time which one has lived through. I distinctly feel now that it was virtually yesterday that I was writing this book, even though several years separate me from those days and so much has happened and so much has changed. In that period, I began working in a new place, as astrophysics professor of Copenhagen University. My native country, the former USSR, the former enormous empire, broke into pieces and is trying, in untold hardship for its peoples, to claw its way out of the frightening historical abyss into which it had been plunged. Even though I continue to head the Department of Theoretical Astrophysics of the Petr Lebedev Physics Institute in Moscow, my settled life beyond the borders of my native land, in a very different world, has definitely changed my perception of life, although to a considerably lesser degree than I could have predicted. It involves my attitude to this book as well.

My recollections of childhood days that are found in the preface to the Russian edition are likely to be more understandable to the Western reader if I now add several strokes to this description; I do think that these are rather typical for my generation in Russia. I have mentioned already in the preface that my father fell victim to Stalin's regime when I was two years old. I do not remember anything about him. My mother, arrested and exiled, ultimately returned from the Gulag areas but was not allowed to live in Moscow. She secretly visited my elder brother and me in a tiny 'communal' (multi-family) three-room flat occupied by my stepfather, my grandmother, my brother's wife and the family of my aunt (four people, including my cousin suffering from tuberculosis). My mother was terribly tormented by the utterly unexplainable and meaningless persecutions of Stalin's system. Not only was she, a very beautiful young woman, snatched out of life in the 1930s and thrown into the hell of Gulag prisons but, later, she was constantly trying to comprehend – and failing to – 'What was it for? What have I done?' Constantly remembering how my father had been arrested and then herself, she got so terrified in the nights by a noise or a knock on the door that she would throw herself under the bed, with a hysterical, barely audible yell: 'They've come... they've come to take me!'. My brother and I were marked with this invisible brand of 'children of an enemy of the people'. Those who were never branded with this secret, destructive, caused-by-nothing label which put you beyond the pale of law and society could not be expected to grasp the weight of this load.

I need to remark that nobody in the family ever displayed any hate towards the reigning political system or even betrayed a critical discussion of it, at any rate in the presence of children. Perhaps, the adults suffered so much grief that they shielded their young from the reality. I think now that I simply had no idea that other ways of life, other surroundings were at all possible, and thus could not suffer 'excessively'. I regarded even our utter poverty as something that was to be taken for granted. When my stepfather died, leaving mother and me to live alone, our monthly budget was about 600 roubles, the price of lunch in a student canteen being 8–10 roubles. My brother would help as much as he could: he began to work. But it was very little: he had his own family to feed on an engineer's salary, which was quite modest in the USSR.

My early passion with the mysteries of the Universe was invariably encouraged by my relatives. I would switch to a different world which was far removed from the all-pervasive tragedies of my country (so I was hardly conscious of them), to the world of pure truths devoid of the contradictions of our day-to-day existence; I fell in love with the logic of relations between these pure truths. This may have been too deep a devotion, since, from the earliest moments that I can recollect about my childhood and youth, I was absolutely sure that the most important and deeply loved truths about space, time and the Universe had at last (and only recently!) been understood and established as final. I did not discern (or tried not to notice) the obvious discrepancy: my attitude meant that the millennia-old history of science had timed the discovery of the most important knowledge about the world almost to the day of my birth. Having become a scientist, I had to fight in myself this extremely harmful and unproductive attitude of a person who believes that he knows – or can find – the ultimate truth. In fact, such beliefs are dangerous, and not only in science but in life as well.

These were therefore the psychological surroundings in which grew my love of Knowledge, which I perceived as the love of the Grandiose, Mysterious (especially Mysterious) and Eternal.

The science of astronomy existed in my country in the rigid, draconian reality warped by Stalin and his henchmen. Officially, all science was classified into two groups: the 'progressive, the-only-true, *our* Marxist science' and the 'decaying, on-the-brink-of-bankruptcy, *their* capitalist science'. Today in Russia, as always in the West, this sounds as a flat hoary joke. The reality was far from a joke, it enforced a form of existence on science. The theory of an expanding Universe was banned. My future professor and advisor Abram Leonidovich Zelmanov, a cosmologist, one of the creators of the mathematical apparatus of today's science of the Universe, was fired, together with some other leading scientists, from his job at the Shternberg Astronomical Institute in Moscow: both for his research in cosmology and for being a Jew. I vividly remember how, still very young, I pounced impatiently on a fresh issue of the recently organized *Referativny Zhurnal* for abstracts of the latest papers on cosmology in foreign journals, only to be stunned by a cliché at the end of each one of them: 'The author (or authors) shares the views of the bourgeois theory of expanding Universe'.

In our country of that time, scientists had to think – first and foremost – about survival, at the same time doing the job they were devoted to. It is a marvel, perhaps, that in such an atmosphere the science of the Universe did not degenerate in the country; in fact, it even produced exceptionally good results. I tend more and more to the opinion that the 'double burden' on the shoulders of our experts, in some way stimulated a successful quest for new knowledge. It made them work with quadrupled effort and yield.

As a tentative proof of the possibility of such a response to crushing calamity, I turn to a contemporary genius in astronomy and physics: Stephen Hawking of Cambridge University in England. Hawking has been crippled by a frightful disease and confined to a wheelchair; with time, he has virtually lost all control over his muscles and finally lost speech. His intellect and his sense of humor, however, were getting stronger and sharper. As one of my colleagues put it, Hawking was transferred to a different life dimension and there achieved outstanding results in science.

I do regard myself as an expert in, among other fields, the physics of time; hence, it is impossible to forget, in a book about time, to mention some moments of my personal experience of 'floating' in time, of being carried by time flow and its vortices through the middle and end of the 20th century.

Any person who devotes enough thought to the meaning of 'being' comes, sooner or later, to query the very hypothesis of 'climbing on the banks of the River of Time', of liberating oneself from its majestic flow, of stopping and, so to speak, looking at the essence of what is happening.

The query will cease to appear so strange if one remembers that we are indeed able to stop traveling through space and 'come to rest'. Why then are we unable to do this in time? Or are we?

However, I jump ahead of the story here – more of that later in the book.

The book has been substantially revised for the English edition. Some passages, which seemed to overload the presentation, have been dropped. On the other hand, I have added new material, mostly dealing with further progress in the analysis of the possibilities of creating a time machine, paragraphs outlining some of my discussions with colleagues in Russia and in the West, and much more.

To conclude this 'second' preface, I wish to add several words to the comparison, offered in the 'first' preface, between art, on the one hand, and talking about science, on the other.

One can rather crudely divide painting into, say, realistic and abstract. Both types of painting stimulate feelings and thoughts in the viewer (profound feelings and thoughts if the paintings are truly great). However, abstract art requires that the viewer participate in the process of creation of a painting, that he 'think further' and 'feel further' into what the painter has presented. Realistic painting generates very different associations stemming from the comprehensive visual images that the painter has completed to perfection.

I believe that a story about science (my story, anyway) is closer to realistic than to abstract painting. I do not exclude that one could write about science in the 'abstract-art' style, inviting the reader, who is not an expert in the field, to join in drawing the conclusions. The fantasies and dreams of a non-expert reader may then carry him or her too far astray. This might be interesting, may even be desirable (I might try and write something like this some day) but this would definitely *not* give a picture of the current status of a science. Science is not a dream but a reality, often very useful, a practical and necessary reality. I do not forget, of course, that without a dream, one can never achieve important results in science.

For the English edition I added a few new illustrations partly using the characters of the illustrations in the Russian edition.

And last but not least: in recent years during the preparation of the English version of the book, in addition to my duties as a Professor of the Astronomical Observatory of Copenhagen University, I also worked as Director of the Theoretical Astrophysics Center of the Danish National Research Foundation. Both institutions supported and encouraged me in my work, and I thank them very much for that.

I. D. Novikov
Copenhagen

Origins of thinking about time

Ever since I started reading popular science books on physics, I have regarded it as self-evident that time is synonymous with empty duration, that it flows like a river and carries in this flow all events without exception. This stream is unalterable and unstoppable, going in a never-changing direction: from the past to the future.

It seemed that this interpretation, given our knowledge about the surrounding world, was unavoidable.

I learnt only many years later that people had not always held such or similar intuitive notions – far from it.

Heraclitus of Ephesus, a philosopher in ancient Greece who lived at the end of the 6th century BC, appears to have been one of the first thinkers of antiquity who set forth a belief that everything in the world changes and that this changeability is the highest law of nature *(all things are in process and nothing stays still)*. Heraclitus set out his view in the book *About Nature*, of which only a few fragments survived and reached us (*Cosmic Fragments*).

Heraclitus taught that the world is full of contradictions and variability. All things undergo changes. Time flows relentlessly, and everything that exists moves with this unstoppable stream. The skies move, physical bodies move, a human's feelings and conscience move as well. 'You cannot enter twice into one and the same river' said he, 'because its water is constantly renewed.' Things come to replace other things. 'The fire is alive through the death of the earth, the air is alive through the death of the fire, the water is alive through the death of the air, the earth is alive through the death of the water.'

From the high ground of our current knowledge, we tend to look down with irony on the chain of births and annihilations described by Heraclitus. Nevertheless, he gave a very impressive picture of the general changeability of all things in time: '... everything is changing in the all-encompassing circulation in the creative game of the Eternity'.

Science was just emerging in those distant centuries. The thinkers of that period had not yet formed the concept of directed progressive development. People rather observed the cyclic organization of phenomena in the surrounding nature. Day was replaced with night, to return again in the morning. One season was replaced with the next and was resumed at the end of the annual cycle. The motion of heavenly sources of light was cyclic too.

As a result of these constantly observed phenomena, time was not perceived as an omnipresent unidirectional flow – as a 'river of time'. Time was pictured rather as a cyclic alternation of opposites. For instance, the Greek mathematician and philosopher Anaximander of Miletus (c. 610–547 BC) taught that the primal basis of any existence was 'infinity'. Its eternal motion generates the opposites: heat and cold, dryness and moisture; then everything returns to the original state. Anaximander stated:

> The primal essence of the existing objects is also the fact
> that when they perish, they return as dictated by necessity.
> Indeed, they justly reimburse one another in a prescribed
> time as a compensation for damages.

I believe now that this is a very original interpretation of time and changeability, one that relates them to the concepts of justice and balance.

However, the idea of only temporary cyclic changes and the invariance of the totality of the existing world reigned in the minds of thinkers during many centuries. People believed that all phenomena change cyclically, returning 'to their proper orbits'.

The famous idealistic Greek philosopher Plato (427–347 BC) advanced interesting and profound ideas concerning time.

Plato was a pupil of Socrates (470–379 BC), known as 'the wisest of Hellenes (Greeks)'. He belonged to a very rich and old family whose origins can be traced back to the last king of Athens. We know very little about the life of Plato and most other philosophers of that period. Reliable facts are interspersed with legends and even obviously apocryphal anecdotes. We know that Plato received a complete course of training under the guidance of the best teachers. This means that he studied grammar, music and gymnastics. Then he began to write poetry. In 407 BC, the twenty-year-old Plato met Socrates and devoted himself completely to philosophy.

Socrates' method of teaching was to conduct a free discussion with anyone willing to listen to him. The rulers banned such talks with young pupils but the philosopher held to his principles and patriotism and ignored the orders. His unfettered disputes with pupils had a tragic finale: he was accused of godlessness and perverting the young, and incarcerated. Friends offered help to escape from jail, pupils (including Plato) collected money for bail. However, Socrates chose a proud line: he rejected running from prison, was given a harsh sentence and had to drink a phial of poison.

After his teacher's death, Plato moved to Megara and continued studying philosophy. He traveled extensively, attempting to influence the rulers into creating an 'ideal state' run by philosophers. These attempts failed utterly. Some (unreliable) evidence claimed that he had even been sold into slavery but freed himself and returned to Greece. Having regained Athens in 386 BC, Plato founded his school of philosophy that he called the 'Academy'.

Plato taught that the world that people observe and study is not the 'true world' but is merely its external incarnation. Both heavenly bodies and bodies on the Earth are but 'pale shadows' of some ideal objects which constitute the true world: 'These shadows are imperfect and changeable'. Plato taught that the 'true world' consists of abstract essentials (he called them 'ideas'). The 'ideas', these 'spiritual entities', are impeccably perfect and unchangeable in principle. Ideas exist not in our material Universe, not in space and time, but in the ideal world of complete perfection and eternity.

The true existence, said Plato, is the ideal existence. For example, the true abstract world includes not a specific thing, say, a wooden table of certain color, shape etc. but the abstract notion of 'table'. This notion is the 'idea of the table'.

Obviously, this idea cannot change. The eternal unchangeability of ideas resembles the properties of geometric figures: triangles, circles, pyramids. Their properties also remain absolute, they also exist in the abstract world of one's mind. However, Plato postulated that the true reality was this abstract world.

According to Plato, the Creator (Demiurge) conjured up the visible world by 'copying' these ideal objects. Each body tends to resemble the original but is inevitably changeable, has a beginning and an end. As a result, the 'pale shadows' fail to reproduce their ideals. The ideals personify eternity, while the world as we see it constitutes constant changeability. To put things in order and to smoothen the contradiction, the Demiurge devised time. 'His idea was to produce a non-static resemblance of eternity: while arranging the heavens, He created for the eternity (which stays unified) an equally eternal reflection which moves from a number to a number that we call time.'

Therefore, by analogy with the bodies in the surrounding world that we perceive by vision and touch and which are, according to Plato, imperfect copies of their ideal originals in the world of ideas, time is an imperfect 'model', an image of the ideal eternity. Time is perpetually flowing, thus imitating the unchangeable perfect abstract eternity of the abstract world of ideas.

This sounded very beautiful. Plato even thought up a mechanism for time to arise in the world created by God. Time, he said, is born in the motion of heavenly bodies, in the perpetual and unchangeable cyclic motion of the Sun, the Moon and the planets that man observes. In fact, Plato identified time with this cyclic motion.

Since the motion of heavenly bodies is cyclic, time also appeared to be cyclic, running on a circle. According to Plato, everything in our world repeats itself after a large segment of time. (Plato even indicated the length of this period: 36 thousand years.)

So many centuries separate us from antiquity that it is often very difficult to realize the level of knowledge of that time and the style of reasoning typical of that culture. It is therefore often almost impossible to appreciate the true measure of the scientific genius of a thinker in antiquity who made a bold step on the infinitely long road to uncovering the truth. For these reasons, and also owing to the paucity of reliable data, it is even more difficult now to reconstruct the complicated, multifaceted personalities of the philosophers, their far-from-simple life stories.

At that period, sciences were not divided cleanly into branches, no science could be distinguished from the all-encompassing philosophy, psychology and ethics. The knowledge, the feelings, the social and ethical positions often intertwined and affected one another. Plato chose for his writings the form of dialogues; in all likelihood, they were not a systematic presentation of his ideas, meant to follow a previously thought-out plan. The dialogues were written at different periods of Plato's life and at least some of them were stimulated by his debates with sophists (who preached intellectual anarchy) or other opponents, and by various problems in his life. The dialogue in these debates is always led by Socrates.

Plato's points of view changed with time. While still a pupil of Socrates, he believed that a philosopher lives to achieve cognition of abstract truths by way of free exercise of mental power. This cognition leads to happiness and is independent of external circumstances. Following Socrates, he postulated that the evil in the world stems from the ignorance of people, from their separation from the truths.

The death sentence on the obviously innocent Socrates had shaken Plato, and his outlook changed. He came to the conclusion that a world with this unbearable amount of malice cannot be the true one. The true world is the realm of perfect ideas. In this period, Plato was very skeptical about the view that teaching people what is good was the philosophers' goal. He believed that people were incorrigible. In one of his dialogues he painted a portrait of the principal accuser of Socrates. This anti-hero proclaimed that only government employees were true teachers of good while the so-called sages were merely malicious saboteurs of the foundations of society. In this dialogue, Socrates asked whether the anti-hero was acquainted with sages, and received the answer that no, he was not, nor would he wish to be, but that he strived to inflict as much harm on them as possible...

At later stages of his life, Plato tried to create in his writings a model of a state that he would consider as the 'ideal' one, the state ruled by philosophers; in fact, this was described as a state with slavery and wars, where Greeks were placed unquestionably above all the others (the barbarians). Later Plato made attempts to actually change the social structure by influencing the rulers, but in this, as I have already mentioned, he failed completely.

In his last book, *The Laws*, which Plato most probably wrote in his old age, he totally reneged on the striving of his younger years for truth and fairness. This is one treatise of Plato where the shining image of Socrates is not central at all; in fact, the teacher is not even mentioned. The spirit of this work is completely opposite to Socrates' principles.

The Codex of Laws that Plato compiled for the future 'ideal state' on Crete incorporated criminal persecution of 'magicians', the death sentence to a slave who failed to inform the authorities about a 'violation of social serenity', the death sentence to anyone who would dare to criticize the social order protected by the authorities and the official religion. In this way, Plato at the end of his life, shifted to the position of Socrates' accuser, anti-hero, whom Plato had attacked before.

Plato was one of the greatest thinkers. Later generations tend to idealize the image of a great man. However, even great personalities are not always entirely consistent. More often than not, they are complicated and contradictory, and change in response to external influences. They are simply human. A well-known German philologist and expert on Plato's work G. Ast (1778–1841), driven by the noblest intentions, went out of his way trying to classify *The Laws* as a fake text which is only attributed to Plato. Alas, this is most likely Plato's true work. We have for this the words of Aristotle of Stagira (384–322 BC), who was the most famous of Plato's pupils. The contradictions we find in Plato, his very complicated life and the undisguised reactionary nature of some of his statements, by no means detract from his tremendous contribution to science and philosophy.

Let us return to the problem of time. Aristotle held the same view on the cyclic nature of time as Plato, his teacher, had. Aristotle, one of the greatest scientists of ancient Greece, was an illustrious personality. His father was a doctor at the royal court of the Macedonian king. The father taught his son medical subjects and philosophy, wishing for Aristotle to inherit his position at court. Life changed these plans drastically. Having lost both parents quite early, the eighteen-year-old Aristotle went to Athens and entered Plato's Academy. He very soon mastered the philosophy of his teacher and rose to an independent position. His views deviated considerably from those of his teacher. Immediately after Plato had died, Aristotle left Athens. In 343 BC the Macedonian king, Philip, entrusted Aristotle with the education of his son Alexander, the future famous commander, Emperor Alexander the Great. The ennobling influence of Aristotle must have been quite strong, despite the atmosphere of plots and intrigues that reigned in the royal palace. Philip and Alexander felt immense gratitude to Aristotle, richly rewarded his services and rebuilt the ruined Stagira, his native city. Later various conspiracies destroyed the friendly relationship between Alexander and Aristotle.

But before that, Aristotle had returned to Athens in 334 BC and founded his own school, known as the Lyceum of the Peripatetic. The name of the school may have been connected with Aristotle's habit of constantly walking during his lectures.

After Alexander's death, the party of Independent Greece resisted Macedonian rulers and thus regarded the former teacher of Alexander the Great as a dangerous influence; furthermore, Aristotle enjoyed great respect from the young generation surrounding him. Aristotle was thus accused of godlessness; this stratagem was used against scientists by their enemies both before and also many centuries after Aristotle's time, and proved to be very convenient, since it was easily accepted by the ignorant populace. Aristotle realized that a just trial was impossible and that he would share Socrates' fate unless he decided to flee. He did leave Athens at the age of 62 and died fairly soon after.

It appears from the remarks of his contemporaries that Aristotle had a sarcastic wit and a very sharp tongue. His witty speeches were meant to ridicule his opponent, and he was cool and jocular. If we add to this that he was short, wizened, short-sighted and had a lisp, we can easily imagine that he had no difficulties in creating enemies.

It appears that Aristotle made no attempt to be delicate in his arguments and in demonstrating the power of his reasoning. We do not know whether this behavior was deliberate or unconscious. Incidentally, many centuries later another genius – Sir Isaac Newton – formulated at a relatively young age (he was 27 at the time) a different principle which stated, roughly, that needlessly parading one's superior intellect would only harm whatever work was being undertaken. He wrote in a letter to an acquaintance in Cambridge that you stand to gain little or nothing by appearing to be wiser or less ignorant than the society around you.

Perhaps, these very different attitudes towards social relations reflect not the distance of thousands of years but the differences in temperament and, generally, the fact that just as ordinary people, all geniuses are vastly different.

Aristotle left a tremendous imprint on all later developments in science and philosophy. His writings gave a summary of the current status of the whole of science and greatly contributed to some of its fields. In contrast to Plato, Aristotle rejected the notion of a non-material time-independent world of ideas. He believed that the world that we observe by vision and touch was real. Aristotle regarded physics as the science treating changing objects that exist in the real world. This distinguishes physics from mathematics which studies inherent, unchanging properties of numbers and shapes. Nonetheless, his physics remained a contemplative science.

According to Aristotle, the primary properties of matter are opposites: 'warm' and 'cold', 'dry' and 'moist'; the primary elements are earth, air, water and fire. To these he added the most perfect element, the ether. Aristotle taught that the main elements – earth and water – tend to move 'downward', towards the center of the Universe (this was his explanation of weight); we would say that they are subject to a force that pulls them down. Conversely, air and fire tend to rise up (in our language, we would say that a 'lifting force' acts on them). It is of interest that the separation of the contents of the Universe into 'physical matter' and 'interaction forces' survived in physics to our days, even though they came to mean very different things.

Aristotle taught that the Earth was spherical and stationary, and that it was located at the center of the Universe. He taught that the Moon, the Sun and the planets are fixed to crystal spheres and revolve around the Earth along concentric circles. Their motion is driven by the revolution of the outermost (astral) sphere to which the stars made of ether are fixed. The region within the orbit of the Moon ('the under-moon realm') is a region of various nonuniform motions. Everything beyond the orbit of the Moon (the 'above-moon realm') is the region of eternal, uniform, perfect motion.

In contrast to Plato, Aristotle assumed that motion, even the most perfect rotation of the astral sphere, was not yet identical to time. He believed that time makes it possible to measure motion, that 'it is the number of motion', that is, something that allows us to decide whether a body moves fast or slowly, or is at rest.

However, it was not the motion itself, that is, the dynamic process, that interested Aristotle in physics but the previous and subsequent states of the body, the initial and the final states, so to speak. As a result, time did not play for Aristotle the important role it plays in the physics of our time.

In subsequent years Aristotle's teachings were canonized by the church; Aristotle's words were treated as 'uniquely true'; a ban on any changes in his picture of the world became an obstacle to further progress in science.

I will conclude this visit to ancient times by quoting the philosopher Boris Kuznetsov on the culture of that period.

> On the whole, the culture of Antiquity creates a feeling of
> a grandiose turn in the way people thought and felt, of an
> expansion in the scope of concepts, of logical norms and
> factual information that happened in Antiquity. When we
> look at the statue of Venus of Milo, her beauty strikes us
> with an image which is multifaceted, infinitely dimensional
> and at the same time completely harmonious. This
> impression is so intense that it as if encloses in parentheses

11

the entire later progress of civilization; likewise, the childhood of a person fascinates us with its promise, its novelty and its freshness, something that can never be repeated.

2

Science of time is born

The Renaissance that came to replace the somber Medieval centuries brought outstanding discoveries in natural sciences. This was the time when Nicolaus Copernicus (1473–1543) developed his theory which was to produce a dramatic transformation in people's view of the world. First of all, this new concept eliminated the impenetrable barrier between the terrestrial and the celestial. Before, everything celestial was a symbol of perfection, of eternity, and of ideals. Heavenly bodies were ideal, as was their uniform motion along circular orbits. This perfection was in opposition with the rough terrestrial matter and its chaotic irregular motion. Copernicus' model showed the Earth to be an ordinary planet which revolves, just as other planets, around the Sun.

Nicolaus Copernicus became a canon of a Catholic church in Frauenberg [Frombork], a small town on the banks of the Vistula in Poland, in 1510. In quiet solitude, he worked on his astronomy. In fact, he spent his free hours on other things as well. He treated patients for no fee. A new monetary system was introduced in Poland following his proposal. He designed and constructed a hydraulic machine to supply water to households.

Copernicus was very careful about publishing his results; he clearly recognized the contradiction with the church's teaching of the singular position of the Earth and man in the Universe. His treatise, *On the Revolution of Celestial Spheres*, dedicated to Pope Paul III (this was agreed upon with the Holy See) was printed in 1543, not long before Copernicus' death. In fact, Copernicus had formulated his main conclusions long before the publication. He wrote in his work 'Smaller Commentary', dated approximately 1515:

> All the motions we observe as those of the Sun do not
> belong to it but to the Sun and our sphere together with
> which we revolve around the Sun, as any other planet does;
> the Earth thus executes several motions. The apparent direct
> and reverse motions of planets are not theirs but of the

Earth. Therefore, this one motion of the Earth is sufficient to explain a large number of irregularities observed in the sky.

In our day, it is quite difficult to imagine to what degree a man's way of thinking had to be non-trivial to dare to claim at that time that the Earth was not stationary. The point here lies not only in the disagreement with ecclesiastical dogmas. Indeed, Aristotle's teachings reigned in science, stating that force is constantly required to sustain any motion (science did not yet know anything about motion by inertia). It was assumed, as a result, that if the Earth revolved, this would affect terrestrial phenomena: the air would tend to stay behind, thus creating hurricanes on the rotating Earth; a body dropped off a tower would not fall to its foundation since the ground would fly away from under it, and so forth.

This shows that Copernicus had to argue mostly against the Aristotelian misunderstanding of motion which was rooted in a long chain of centuries before him. There was another reason why these wrong notions were so difficult to overcome. Namely, people thought that no observations or experiments were needed to obtain knowledge about nature: it would be sufficient to think hard and reason by logical inference for the truth to be established.

Using astronomical observations, Copernicus had not only created a new model of the Solar System but was in fact the first to challenge the dogmas of Aristotelian physics. He understood that everything on the Earth moving by inertia must occur exactly as it would on the Earth at rest:

> Why not assign the appearance of the daily rotation to the sky and the reality of it to the Earth? Indeed, when a ship sails on a quiet water, everything outside the ship appears to the seamen as if moving in accordance with the motion of the ship, while they and everything with them on board the ship appear to be non-moving. The same can undoubtedly take place on the moving Earth and one may

15

conclude that the whole Universe is rotating. What should we say then about the clouds and everything else which somehow hovers, descends or ascends in the air? Only that not only the dry land moves together with the water spaces connected to it but also with a considerable part of the air and everything which is in some way connected with the Earth...

... For this reason, the air contiguous to the Earth and all things hovering in it must appear as quiet to us, provided it is not driven now in one direction and then in another, as often happens, by winds or by any other external force.

This passage clearly characterizes the relativity of motion and the properties of motion by inertia, whose final formulation was given by Galileo a century later.

It is quite likely that any person who first learns the laws of mechanics in childhood or youth, has to make a considerable conscious effort to digest the notion that an object dropped from some height in a moving windowless carriage falls to one's feet exactly as it does in a stationary one. In our time, with its frequent travel by train, car or airplane, one gets used to this from childhood. I distinctly remember, nevertheless, my amazement at the age of ten in a truck that was running fast in the Kherson steppe. I watched a ball falling from my hand, again and again, exactly to a point on the floor right below, even though the speed of the truck was huge – by the standards of my childhood. I imagined that the floor of the truck would rush away from under the ball. It was not easy to comprehend that the ball released from my hand continued to move by inertia along with the truck and retained the same velocity that it had in my fist, the velocity of my body and of the truck, before I let it go.

At the beginning, Copernicus' teaching did not cause any special worry for the Catholic Church. The impact was partly cushioned by the unsigned foreword to *On the Revolution of Celestial Spheres*, written by an anonymous theologian. It claimed that the author only aimed at offering a method of mathematical calculation of the observed positions of heavenly bodies and in no way attempted to determine the actual motion of these bodies. It said: 'His hypotheses may be wrong, may even be improbable, as long as they lead to calculations that fit our observations.'

However, at the beginning of the 17th century when Copernicus' theory began to spread in Europe as actual rejection of the dogmas of the church, the treatise was placed into the 'Index of Banned Writings' where it stayed for more than two centuries.

In this period, Galileo Galilei (1564–1642) developed a new understanding of physics, and formulated the first truly substantiated foundations of the science of time, which were later beautifully developed in the work of Isaac Newton.

Galileo made a great many important discoveries in science that the reader undoubtedly knows about. However, the most important of these was his novel approach to natural sciences, his belief that to study nature one has first of all to set up carefully thought-out experiments. The world around us can only be understood by testing a hypothesis in experiments, by 'asking questions of Nature'. Here he parted ways sharply with Aristotle, who assumed that the world could be understood by purely logical reasoning. Galileo also believed that superficial observations not accompanied by thorough analysis of data can lead to wrong conclusions.

Taken together, this was the beginning of the modern method of studying nature. Einstein said that 'the science relating the theory and experiment was actually born in Galileo's work'.

Galileo's discoveries in physics were based on numerous experiments that he had conducted. Especially important for our story is the discovery of inertia and inertial motion.

Everyday observations on the motion of bodies for many centuries had convinced people that unless the motion is sustained, for example by pushing a rolling ball, the body will stop. Aristotle summarized these observations in the following form: 'A moving body will cease to move if the force pushing it ceases to act on the body.' We know now that the rolling ball stops not because no force keeps pushing it but because it is slowed down by the force of friction connected with surface roughness and air resistance. If the surface is made gradually smoother and flatter and air is removed, the ball will roll farther and farther. In the limit, it may not stop at all. This was Galileo's conclusion: '... horizontal motion is eternal since if it is uniform, then nothing weakens it, or slows it down, or destroys.'

The law of motion by inertia discovered by Galileo is the basis of the principle of mechanical relativity. This principle states, for example, that regardless of whether a ship is at rest or sails at a uniform speed on smooth sea, all processes in a cabin proceed identically. One can walk, one can drop objects, flies can fly freely throughout the air, and the motion of the ship has nothing to do with this. Here are the words of Salviati, one of the protagonists in Galileo's book *Dialogue Concerning the Two Chief World Systems – Ptolemaic and Copernican*:

> Lock yourself with a friend in the stateroom under the deck of a large ship, having brought with you flies, butterflies and other small flying animals. Take with you a large fish tank with fish swimming in it. Suspend a bottle from which water drips, drop by drop, into a wide vessel underneath. As long as your ship does not move, watch carefully how the insects fly through the room at the same velocities in all directions. Fish swim randomly, without preference to any direction. Drops fall into the vessel under the bottle. If you throw anything to your friend, your effort will be the same no matter in what direction you threw it, provided the distances are identical. If you jump pushing with two feet at the same time, you cover the same distance in any direction.

Having carefully observed all this (even though you never doubted that it would be exactly like this in a ship at rest), order the crew to set the ship in motion at any speed, but so that the progress of the ship is uniform and not disturbed by anything. You will not discover any changes in the motions you were watching and will be unable to identify by any of the processes whether the ship is moving or not. Having jumped, you will cover the same distance as before, and a jump towards the bow will not be shorter than that towards the stern, even though the ship was moving under you while you were in the air, and in the latter case in the opposite direction. To throw an object to your friend, you will not need to spend a greater effort if your friend is closer to the bow than you are. The drops will keep falling into the vessel below as before, without deviating towards the stern, even though the ship moves forward several feet while drops fall through the air. Fish continue swimming in their tank with equal ease in all directions and catch bait in whatever corner we choose to place it. Finally, flies and butterflies are flying in all directions without preference, and you never find them clustering at the stern, as if getting tired to keep up with the progress of the ship from which they were separated, being suspended in the air for a considerable time.

This wonderfully expressive description is one of the first formulations of the principle of relativity of motion. Note that Galileo's writings are not only collections of gems of human thought but also outstanding literary work. Schoolchildren in Italy study them first of all as the literary heritage of their country.

No mechanical experiments inside the stateroom can determine whether the ship is moving or is at rest. I have said already that in our era of incessant car, train and air travel, we became used to this a long time ago. It is instinctively clear to us that a statement 'the cup is at rest' is meaningless unless we specify that it is at rest with respect to another object. The cup may not be moving with respect to us in a flying plane but may move together with us at a high speed with respect to the Earth. We can saunter leisurely through the aisle of an airplane while traveling at a great speed relative to the Earth. As for any motion, the rest state of a body and its velocity are relative; these terms are all meaningful only when we indicate the 'laboratory' with respect to which these notions are being used.

This discovery by Galileo Galilei – namely, that everything proceeds identically, regardless of the uniform motion of the 'laboratory' in which the observations are made – was a scientific argument against the belief that the Earth is at rest in the Universe. Following Copernicus, Galileo stated: 'Let us choose for the foundation of our cognition the concept that whatever be the motion of the Earth, the inhabitants of the Earth do not notice it as long as the judgments are based on things terrestrial.'

Galileo firmly believed that Copernicus' teaching was true and became its passionate propagandist. Galileo's discoveries in physics and astronomy made him the most famous scientist in Europe. At an early stage, the Catholic Church made cautious attempts to cajole Galileo to change to the point of view that Copernicus' model was only a hypothesis convenient for calculations (as Osiander claimed in his foreword to Copernicus' treatise). Cardinal Bellormino wrote to Father Facarini, who sided with the Copernican picture of the world:

> It seems to me that You and Seignior Galileo would make
> a wise and careful move if You chose to be satisfied with
> *suppositione* statements and not insist on absolute ones;
> Copernicus' words, and as I always believed, his thoughts
> agreed with this position. Indeed, when one claims that all
> the phenomena observed are saved better when assuming
> the Earth to be moving and the Sun to be at rest than by

postulating epicycles and epicenters, this claim is very well
formulated and not fraught with any pitfalls; and this is all
mathematics needs; if, however, someone begins to talk of
the Sun as actually being at the center of the world, only
rotating around itself but not journeying from east to west,
and of the Earth as placed on the third heavenly sphere
(being the third closest to the Sun) and moving at high speed
while revolving around the Sun, this is a very dangerous
thing, and not only because it irritates all philosophers and
theology scientists, but also because it harms the Holy
Faith, since it implies that the Holy Scriptures are lying.

The Soviet physicist Vitaly Ginzburg remarked that the benevo-
lent permission to 'save' phenomena and do mathematics but shun
the reality caused Galileo's fury. Galileo wrote in a letter to the
Duchess of Lotharingia:

Professors of theology should not claim the right to
regulate with their decrees such professions that do not fall
under their authority, because you cannot impose on a
natural scientist an opinion about natural phenomena...
We preach the new outlook not to sow confusion in the
minds of people but enlighten them; not to destroy science
but to give it a sound foundation. Our opponents, however,
call everything that they cannot disprove a lie and a heresy.
These philistines make themselves a shield out of their
hypocritical religious ardor and dishonor the Holy
Scriptures by using them as a tool for pursuing their own
end... To prescribe to astronomy professors to use their
own intellect to seek protection against their own
observations and conclusions, as if these were mere
deception and sophisms, would be a demand more than
impossible to meet; it would be the same as ordering these
men not to see what they see, not to understand what is
clear to them, and draw from their studies the conclusions
that are just the opposite of what is obvious for them.

Ginzburg added that these words sound as if written by a contemporary.

It is left for me to emphasize that not all movement of the 'laboratory' is unnoticeable to people and objects inside it, far from it. For instance, if a car accelerates abruptly, or makes a sharp turn, we feel it very distinctly. Only uniform motion along a straight line is unnoticeable. Such motion of a 'laboratory' or a body occurs by inertia, without any forces acting, or when all the 'pushing' and 'resisting' forces, and those forcing a body off the rectilinear trajectory, exactly balance one another out; such motion is known as inertial motion and the 'laboratories' as 'inertial laboratories'.

Of course, a 'laboratory' found in nature can only be inertial to within a greater or smaller degree of approximation. A ship going slightly up and down on gentle waves is obviously not an 'ideal inertial laboratory'. This rocking of the ship is detectable. However, the smaller the accelerations and the smoother the turns, the closer the 'laboratory' is to an inertial one. The Earth's surface is also a mere approximate inertial laboratory. We know, for instance, that it undergoes a circular motion around its axis.

Specially designed experiments can and do detect this. The reader may have observed, or at least heard of, the Foucault pendulum. The pendulum is a heavy object (ball) suspended in a high-ceilinged building on a long string. When the load swings, it tends to preserve the plane in which it moves with respect to the stars. The surface of the Earth, together with the building, performs its diurnal rotation, and we discover that the direction of swing of the pendulum gradually changes with respect to the walls of the building. Such experiments were first set up many years after Galileo's time, in 1851, by the French scientist G. Foucault, who suspended one in the dome of the Panthéon.

But let us return to the 17th century. True knowledge was clearing its path by a passionate struggle with deeply rooted dogmas, with very profound difficulties that nature always erects for humans in search of truth, and finally, with social conflicts involving the interests of numerous groups of people.

Some time after the notorious trial in 1633 which made Galileo 'Inquisition's captive', he published *Discussions and Mathematical Proofs Concerning Two New Sciences....* In this book which presented the foundations of dynamics he wrote: 'This treatise only opens the door to these two new sciences so rich in applications; they will in the future be expanded immeasurably by inquiring minds... one of the sciences concerns an eternal subject, one of a paramount significance in nature.'

A year after Galileo died, another genius was born: Isaac Newton (1642–1727). His work completed the creation of classical physics and also of the first physical theory of time (in the sense acceptable to us).

In contrast to the lives of philosophers of antiquity, we know Newton's life rather well. At first glance, it was strikingly meager in events. Beginning his story of Isaac Newton, Boris Kuznetsov remarked: 'There was no family, no voyages, there were no major changes in his way of life, almost no friends, almost non-existent social activity. To a superficial view, this list is in stark contrast with an unbelievable intensity of the creative path of this thinker, with true tragedies in the cognitive process. Actually, the two sides are in profound harmony.'

Newton was born in the village of Woolsthorpe in Lincolnshire, England, in the family of a yeoman farmer. His father died several months before the son was born. The boy attended the King's School in the small town of Grantham not far from Woolsthorpe and entered Cambridge University at the age of nineteen. Even at this age he was punctilious, inclined to systematization and order. He began as a poor student of Trinity College, one of the most famous in England, graduated in three years and soon developed into a thinker of exceptional genius. In 1669 he became the Lucasian Professor of Mathematics. The Henry Lucas Chair of Mathematics was established in 1663 on the donation of Henry Lucas and still remains one of the most famous and respected chairs of theoretical physics in the world.

Within the very short period of 1665–1667, while staying in his native Woolsthorpe, Newton formulated the basic physical ideas that gave a new impetus to the progress of physics; he published them much later.

During this period, a plague epidemic was raging in England. Newton left Cambridge, where he had just obtained his BA degree, moved back to Woolsthorpe and there spent about eighteen months. He was working hard, trying to improve the precision of glass polishing, designing physical instruments and conducting chemical experiments. At the same time, he was thinking with great intensity about the main problems of physics, astronomy and mathematics. The results of his work were truly fantastic and deserve being called a revelation. Still staying in the village, he formulated the fundamental laws of physics and created the theory of gravitation. According to this theory, the weight which forces a body to fall down onto the ground is identical to the force which sustains cosmic bodies in their orbits; the magnitude of this force decreases in proportion to the inverse square of distance.

Nearing the end of his life, he recalled that he had noticed an apple falling off a branch, which set him thinking about the causes behind the fall of bodies towards the ground. The answer seemed to be well known to anyone: the weight of a body. But what is the weight? Newton concluded that the weight is the force of attraction to the Earth. The same force must extend further away from the Earth, holding the Moon on its orbit and not allowing it to fly away by inertia into cosmic space.

Newton published the exact formulation of the universal law of gravitation much later, in his famous treatise *Philosophae Naturalis Principia Mathematica* (1687), often referred to in short as *Principia.* (In fact, Newton was always very slow in publishing his results, even though he was definitely not indifferent to priority arguments.) Why did he hesitate? It is likely that the main reason was his very different attitude to gaining knowledge, his concept of the stage at which a result can be recognized as the established truth.

If we can briefly describe his attitude in this respect, it could be: try to achieve total order in the knowledge of nature, try to gain knowledge which is accurately supported by experimental data and adequately described by logic and mathematics. These are exactly the requirements that science sets for us today.

As many other great ideas, the theory of gravitation had its precursors. For instance, Giovanni Borelli concluded that there was mutual attraction between all bodies in the Universe; also, he conjectured that as planets revolve around the Sun, its attraction balances out the centrifugal forces that were discovered by Huygens. Another contemporary of Newton, Robert Hooke, came to the conclusion that the force of attraction between bodies is inversely proportional to the squared distance separating the bodies. We believe, nevertheless, that it was Isaac Newton who created the theory of gravitation.

We do give the highest regard to the vision of other researchers but we recognize Newton as the true discoverer. Why? Because he and no one else gave a proof of his constructs. From abstract arguments, Newton made the step to mathematical calculations, to physical experiments and to the interpretation of astronomical observations.

That was the beginning of the new physics.

Later we will discuss how Newton first formulated the most important properties of time, which constitute the main subject of this book. I wish to remark at this juncture that the discovery of the law of universal gravitation was very important not only for the development of celestial mechanics (describing the forces that control the motion of all celestial bodies in the Universe) but also for understanding what sort of phenomenon time is. In fact, this became clear after a considerable period of time – about three hundred years later, in this century, when it was proved that gravitation affects the rate of flow of time. However, let us again return to the 17th century.

While staying in Woolsthorpe in 1665–1667, Isaac Newton was not only occupied by the problems of gravitation; he also worked in mechanics, optics and mathematics, in which he made fundamental discoveries.

In the post-Woolsthorpe period, until the 1680s, he was mostly interested in optics and also in chemical experiments. In the mid-1680s he wrote and published the main accomplishment of his life: the famous *Principia*. This treatise summarized the fruits of thinking in his Woolsthorpe period and the results of the subsequent development of the ideas conceived at that time.

About two decades separated the time of obtaining the main results and the time of their publication! I have mentioned already that Newton was never in a hurry to publish, always striving for the maximum accuracy of all conclusions and to their logical impeccability. The following events happened to provide a stimulus to writing the *Principia*.

Some time at the beginning of the 1680s, three well-known scientists got together in a London café and were eagerly discussing the problems of the motion of planets around the Sun: Edmund Halley, Robert Hooke and Christopher Wren. By that time it was already known that, as established by Kepler's laws, the planets follow elliptical orbits. The problem that attracted the three scientists was whether it is possible to prove, under the assumption that the Sun's gravitational pull decreases in inverse proportion to the squared distance to the planet, that the orbits must indeed be elliptical. They did not know the solution of this problem. Wren suggested that they set up a symbolic prize – a book at a price of 40 shillings – to the person coming up with the solution. On a visit to Cambridge in 1684, Halley described to Newton their café discussion, to which Newton remarked that he had known the solution for quite some time already! After this Halley succeeded in convincing Newton that he must write a book presenting the proof. This was how the *Principia* was born, edited and published by Halley at his own expense. As we know from the reminiscences of Newton's secretary – incidentally, his namesake – Newton's life in the period of creation of the *Principia* was exceptionally intense. He was never seen to rest, never rode a horse, never played skittles, almost never entertained visitors, slept at most five hours a day, and tried to spend as little time taking meals as possible. He was lucky in one respect: lectures took very little of his time since they were so boring that students did not attend them.

I recall now how I was impressed by stories declaring without a shadow of doubt that success in any field of activity is the result mostly – up to 95% – of a capacity of the person for hard work. Ever since that time, I have blindly believed in this maxim, have found numerous confirmations of it in the experience of my friends, and try to persuade my students and colleagues that everyone needs to follow this principle. 'To work well is to work hard' – I believe, this was said by Newton (or another genius).

After the publication of *Principia*, Newton's way of life began slowly to change. He kept working intensely and fruitfully in science but other fields were also becoming important. His social and political activities seriously occupied him. The more widespread anecdote about Newton's life is that, as a Member of Parliament, he made a single speech, requesting to close a window which caused a draught. The message seems to be that Newton was completely absorbed by his research and neglected all other facets of life. This is not very likely. I am inclined to think that he was quite serious about the non-scientific side of life.

Until the end of his life (he lived to be 84) Newton changed very little. He was rather short, somewhat stocky, usually withdrawn and reserved; his appearance was quite ordinary, very much that of a typical Englishman. True, he was far from easy to get along with.

Another side to Newton's personality must be mentioned. He was very religious. In my country – the former Soviet Union – it was a rule to tacitly turn a blind eye to this fact, especially in books for the young, or at least mention it in passing, as something unimportant. Presumably, exposing it seemed to harm the 'atheistic propaganda', even though suppressing or distorting some traits of the personality of a great man seems to me a far greater evil that cannot be justified by any 'well-meant' intentions.

Yes, Isaac Newton believed in God. This was not unusual for his time. For long decades of unlimited supremacy of communist ideology, this did seem strange both to myself and to most of my peers in the USSR. The system treated religious belief as not only inadvisable and a social risk but, according to the official standpoint, indicated a certain flaw of the intellect.

A reader in the West may regard this attitude to religion in the former USSR as more than peculiar. This was definitely not the strangest thing we find in the history of my country in this century. The attitude to religion is just another example of deliberate, merciless and unqualified warping of souls in the soul-numbing bolshevik epoch. A Western reader will definitely see nothing surprising in Newton's religious beliefs. I know a number of outstanding physicists in the West who are also believers; however, this is a topic for a discussion elsewhere. I only wish to mention in this context Einstein's point of view, whose attitude appears to be fairly close to the position of a number of genuinely great thinkers of our time. He wrote that

> the most sublime and profound emotion that may befall a man is the sense of the mysterious. It lies at the basis of religion and of all most deeply running tendencies in art and science. A person who has never gone through these feelings seems to me to be – if not dead – at least blind. The ability to perceive that which is inaccessible to our reasoning, which lurks hidden below our immediate responses, and whose beauty and perfection reaches us only as weak indirect reflection – *this* is the sense of religion. In *this sense*, I am indeed religious.
>
> My Glaubensbekenntnis (1932),
> in F. Herneck, *Albert Einstein*, Berlin, 1967, p. 254

But we should return to Newton. He was doing research in theology and the history of religion. He believed that God gave the 'primary push' to heavenly bodies, after which all motions in the Universe rigorously followed strict physical laws. From time to time, though, God finds it necessary to intrude and correct the grand 'clock of the Universe' if 'irregularity is anticipated'. In his picture of the world, Newton appealed to God each time he was unable to find a scientific explanation of a phenomenon. This was the case with an attempt to explain the origin of the Solar System and the origin of the initial velocities of the planets. The same happened when trying to explain the beginnings of the history of mankind.

Let us now turn to Newton's contribution to the understanding of time and space.

We begin with space. Newton taught that everything happening in the Universe occurs in empty space, which holds in itself all bodies and all processes. In fact, this space can be pictured as a gigantic laboratory room whose walls, ceiling and floor recede to infinity. Newton referred to this 'absolute', unlimited emptiness as 'absolute space'. In *Principia*, he wrote: 'Absolute space of its own nature, without regard to any thing external, remains always similar and immovable'.

In Newton's physics, time is a flow of duration which involves all processes without exception. It is the 'river of time', whose flow is not influenced by anything:

> Absolute, true and mathematical time, of itself, and from its own nature, flows equably without regard to any thing external, and by another name is called duration.
>
> I. Newton *Mathematical Principles of Natural Philosophy*

Newton's picture of the world was thus clear and obvious: the motion of heavenly bodies takes place in time in infinite empty space. Processes in the Universe can be very complex, diverse and tangled, but regardless of their complexity, they do not affect the eternal stage – the space – and the unchangeable flow of time. Newton postulated that neither time nor space can be influenced, hence the attribute 'absolute'. He emphasized the unchangeability of the flow of time in the following manner:

> All motions can be accelerated and retarded, but the time, or equable, progress of absolute time is liable to no change. The duration or perseverance of the existence of things which exist remains the same, whether the motions are swift or slow, or none at all.

Albert Einstein gave a very illustrative description of Newton's concepts: 'The idea of the independent existence of space and time can be expressed like this: If matter vanished, only space and time would remain (a sort of stage on which physical phenomena are acted out)'.

The reader may exclaim at this point that this is all so obvious, simple and clear that surely everyone interprets space and time in this manner!

This remark is justifiable, but only because these concepts follow from the observation of the motion of the bodies that surround us on the Earth, from observing the motion of giant heavenly bodies and from numerous physical experiments. It is because Newtonian physics generalized the entire experience of science with the motion of objects, and because this accumulated experience is mastered by us when we read school textbooks, that we tend to regard the Newtonian concepts of space and time as 'innate' to us.

One should not forget that any experiment is limited in scale, duration etc. In Newton's time, as well as much later, all experiments and all observations involved bodies which, judging by today's knowledge, move rather slowly. Gravitational fields known in Newton's time must be characterized from our standpoint as weak; finally, the energy of processes known then must also be classified as low in comparison with those that physics deals with in our time. In this framework, everything that Newton said about space and time holds and the motion of matter indeed leaves time and space unaffected. We shall see, however, that this 'indifference' of space and time to what happens inside them takes place only while the above constraints are satisfied.

However, this is a subject for later discussions. For the moment, I wish to stress that Newton's theory gave no reason for raising a question about any special properties or structure of time. Time is a uniform 'river' without beginning or end, without 'source' or 'sink', and all events are 'carried' by the river's flow. Time had no other properties but the property of always being of the same duration. The 'absolute time' is identical throughout the Universe.

In Newton's picture of the world, the meaning of the words 'now', 'before' and 'after' is quite clear for any events in the Universe, whether these events occurred at the same point in space or were separated by hundreds of millions of kilometers. If everything is clocked using the same absolute time, then everyone understands, say, the phrase 'A supernova has exploded at this moment in a galaxy in the Triangulum constellation'. Even though this galaxy is awfully far away from us and we shall see the light of this explosion only millions of years after when it ultimately reaches us, this understanding does not stop us from imagining that the explosion did take place 'now', at this moment of the absolute time of the Universe.

The absolute coincidence in time and the time that is common for the entire Universe are possible because according to the Newtonian theory there are signals which travel from one point to another 'instantly', that is, they propagate at infinite velocity. Gravitation is an example of such signals. If the mutual positions of gravitating masses change, gravitational forces between these masses change instantly throughout infinite space.

In this Universe, if masses have shifted somewhere, it is possible to 'be informed' of this event at any great distance. In this situation, the notion of 'now' is impeccably clear. Even though gravitational forces at large distances from gravitating stars become very weak and extremely difficult to measure, this could be regarded as, so to speak, our technical problem. Such technical obstacles cannot cancel the possibility of instantly determining – in principle – that masses have shifted somewhere far away.

Einstein was fascinated with the clarity and simplicity of Newton's picture of the world, and called Newton's time 'the happy childhood of science'. He wrote that for Newton, Nature was as an open book that he read without effort. The concepts that Newton used to specify his data seem to follow naturally from human experience and from wonderful experiments that Newton described in numerous details and arranged in careful order as precious toys.

In fact, this sunny picture did have a slight 'cloud' which obviously troubled Newton. The point was that no mechanical experiment could detect whether a body is in motion or at rest in this empty space. Indeed, we remember that all processes in a cabin of a ship occur identically whether the ship is stationary or moving. Isn't it really strange that absolute space is there but a linear motion with respect to it can not be measured? This is a flagrant 'ugly', or un-aesthetic facet of the theory.

As our story unfolds, it will be clear that attempts to chase away this 'ugly' cloud finally led to fundamental discoveries in physics a few centuries later.

It should be mentioned that Newton's views on space and time were not the only ones held in Newton's time. Of special interest are the beliefs of the famous German philosopher Gottfried Wilhelm von Leibnitz, Newton's contemporary. Leibnitz worked not only in philosophy but also in physics, mathematics, history, natural law, historical jurisprudence, theology and diplomacy. The unparalleled scope of his interest was at the same time a cause of a certain patchiness of his scientific results. He discovered new approaches, he pioneered novel ideas but rarely pursued these paths thoroughly, bringing them to logical and detailed completion. He tried to integrate most different beliefs of his time and resolve all disputes and contradictions. Leibnitz dreamed about a peaceful accord of science and religion, catholicism and protestantism; he tried to make science international and even to work out a universal world language. On his initiative, the Academy of Science in Berlin was founded in 1700, of which he became the first president. He worked much to help found academies in Vienna and Dresden; he met the Russian tsar, Peter the Great, with whom he discussed the way to plant the seeds of scientific research in Russia and the measures needed to organize the St Petersburg Academy of Science.

This great scientist rejected Newton's absolute space. He maintained that the space is merely a manifestation of an order in the existence of objects and phenomena, that nature has no absolute space free of physical bodies. Leibnitz concluded that space was relative. In the same vein, he rejected absolute time which would flow irrespective of physical properties; he taught that the world is described by a sequence of phenomena following one after another and this is what people call time.

Once, during a joint work with German colleagues at the Central Astrophysics Institute in Potsdam, I had a long talk with the deputy director of the Institute, Professor D. Libscher; we discussed the general properties of time in the light of the discovery of black holes and their fantastic characteristics. Professor D. Libscher drew my attention to the surprising closeness between some predictions of Leibnitz, made three centuries ago, and our current understanding of time. It seemed especially impressive that Leibnitz insisted that there is simply no such thing as the absolute time introduced by Newton. Leibnitz developed a sort of theory on the relativity of time, space and motion. As a result, Libscher and I wrote an article about time in black holes for the Russian journal *Nature* (no. 4, 1985) and I refer those who are interested in a more systematic presentation to this publication.

Having formulated these intriguing arguments, however, Leibnitz went no further; at the time, he was unable to construct a concrete physical theory based on his philosophical thesis. In contrast, Newton's understanding stemmed from a stringent physical theory that he had developed. This theory was the foundation of mechanics, and mechanics was the scientific platform for the industrial revolution to come. Newton's point of view has prevailed.

Newton's physics withstood the test of time. Physics as we know it today has pushed the limits to which the Universe can be scrutinized much further than was possible in his time. Our image of space and time has become much more profound and multifaceted. However, as we have mentioned already, the science of today does not sweep aside anything that Newton accomplished. The properties of space and time and the laws of physical motion that he established for the scope of phenomena apparent to him remain and will remain valid.

However, we can now reach phenomena that Newton could not investigate; they open for us the previously unknown laws of nature and unanticipated properties of space and time.

To conclude this section, it is necessary to mention another very important property of time, first emphasized by the philosopher John Locke with whom Newton was acquainted and who was greatly influenced by the new physics. The property is that the mathematical image of time is a straight line. In contrast to space which is three-dimensional (its three dimensions are length, width and height), time is one-dimensional, formed by a sequence of events following one another.

This image of time as a mathematical straight line proved to be very important for the further evolution of our picture of the world.

3

Light

I was not quite correct when saying that only motion at relatively modest velocities was known in Isaac Newton's time. Of course, this would be true if only the motion of physical bodies was meant. However, from time immemorial mankind knew a process which propagates at a truly fantastic speed. I mean light. What is it?

Suggestions that light consists of particles which are emitted by a glowing body were made in ancient Greece. Aristotle held this opinion and Newton also shared this point of view. Aristotle assumed the velocity of light propagation to be infinitely high. The same point of view was prevalent until the middle of the 17th century. This belief was shared by the great scientists Johannes Kepler, René Déscartes and others. Galileo was the first to attempt an experimental determination of the speed of light in 1688. He placed two torches on top of two hills at a distance of less than one mile from each other. First the shutter of one torch was opened and when the beam of light reached the observer at the other hill, the latter opened the shutter of his torch. The observer with the first torch was to measure the time between the opening of its shutter and the moment when he saw the flash of the second torch. This was meant to measure the time of travel of light to the second hill and back again.

However, no delay was found in these experiments, so that Galileo concluded that if light 'does not propagate instantaneously, then it does so at a tremendously high speed'. Obviously, such a fast motion could not be measured with the devices available to Galileo.

The Danish astronomer Ole Roemer (1644–1710) was the first to really measure the velocity of light in 1676. This is how it happened. In the middle of the 17th century, the Italian astronomer Giovanni Cassini, who became famous for his high-precision observations of planets through large telescopes, compiled tables of the motion of the Jovian satellites discovered earlier by Galileo. Further studies demonstrated that the calculated moments at which the innermost satellite of Jupiter, Io, entered the shadow of the huge planet did not always coincide with observational data. In the month when the Earth, moving around the Sun, was at its maximum distance from Jupiter, the moments of eclipses were delayed in comparison with the calculated values by almost 22 minutes. When the observations were conducted at the minimum distance between the Earth and Jupiter, there was no delay.

When Roemer heard about this, he explained the delay in 1676 by suggesting that light needs 22 minutes to travel along the diameter of the Earth's orbit. By that time this diameter was known with considerable accuracy. Having divided the length of this diameter by 22 minutes, Roemer came up with the first numerical estimate of the velocity of light: about 214 000 km/s. It was found later that the velocity that Roemer reported was less than the true value by about one third.

It was thus shown for the first time that light does not propagate instantaneously: its velocity is finite, even though very high. Only in the middle of the 19th century was the velocity of light measured not by astronomical observations but directly in experiments on the Earth. These experiments, which were in fact greatly modernized versions of Galileo's experiments, were carried out by the French scientists Fizeau, Foucault and Carnot. Their experiments, carried out at different periods and with gradually better and better accuracy, yielded the velocity of light close to 300 000 km/s. At the end of the 1870s, the problem of measuring the velocity of light attracted the outstanding American experimental physicist Albert Michelson (1852–1931). The experiments he carried out at that time gave the value of 299 910 km/s.

This problem attracted Michelson until the end of his life. It was gradually becoming clearer that the velocity of light plays a fundamental role in the structure of the laws that reign in our world.

The final series of experiments to measure the velocity of light in Michelson's laboratory started in 1929. His daughter recalled that in May 1931, in the last days of his life, Michelson, world-renowned physicist and Nobel prize winner, waited impatiently for the final results of his experiments:

> On the seventh of May, Pease (Michelson's assistant) came to Michelson with the latest figures for the new determination of the speed of light: 299,774 kilometers per second. Michelson's face lighted up with an almost child-like pleasure. Knowing that he did not have long to live, he told Pease to pull up a chair and open a notebook at once so that he might start dictation. "Measurement of the Velocity of Light in a Partial Vacuum." The effort exhausted him, and after dictating the first paragraph, he fell into a peaceful sleep...
> On the morning of May 9, 1931, Michelson died.
>
> Dorothy Michelson Livingston, *The Master of Light*,
> 1973 (Charles Scribner's Sons)

These lines are evidence of what sort of people belong to the cohort for whom to gather knowledge about the Universe is the meaning of their lives; due to these scientists, we have penetrated profoundly into nature's mysteries. The current value of the velocity of light, determined by using an atomic clock, is 299 792.458 km/s. The possible error of this value does not exceed 0.2 m/s.

Michelson's name is also inseparable from the experiments which lead to the development of relativity theory. This theory, created by Albert Einstein at the beginning of our century, made it possible to look at the properties of space and time from a completely new standpoint.

Before describing Michelson's experiments, let us step back a century, to the time when physicists tried to figure out the nature of light.

The idea that light is of wave nature was first suggested by the Czech scientist Jan Marzi in 1648. However, a consistent theory of light was only created thirty years later by the Dutch physicist Christian Huygens. This theory explained elegantly a large number of effects in the reflection of light by plates, the formation of moiré films and other interference, diffraction and polarization phenomena, that the corpuscular theory of light was able to interpret only under very artificial assumptions or failed to explain at all.

Physicists had to argue, however, that if light is a wave phenomenon, the waves must propagate through some medium. The reigning hypothesis was that the propagation medium for light waves was the ether: the finest, all-permeating medium filling the entire Universe.

By the end of the 19th century, the theory of light waves propagating through the world ether was gaining ever increasing recognition.

Unfortunately, mind-boggling properties had to be ascribed to the ether. This medium had to possess hugely greater elasticity than ordinary matter, because only then could light vibrations propagate through it at the enormous velocity that we observe. It had to possess perfect zero viscosity, to allow heavenly bodies to move through it without any resistance, which was another feature observed experimentally.

However, difficulties of this sort were easily waved away: indeed, the ether was not 'ordinary matter'. Thus the well known British scientist Thomas Young wrote at the beginning of the 19th century, that in addition to the so-called solid, liquid and gaseous forms of matter, we also know semi-material forms that produce the phenomena of electricity and magnetism, and also ether.

Today's reader may be interested to know that Thomas Young, one of the creators of the wave theory of light, was a uniquely gifted person. He learned to read fluently when two years old and two years later was reciting numerous memorized verses; when eight years old, he had already constructed physical instruments, then rapidly mastered differential calculus and a large number of languages, among which were Greek, Arabic and Latin. He worked as a doctor, a physicist and an astronomer, but by the end of his life he was compiling an Egyptian dictionary.

Young carried out numerous experiments which proved the wave nature of light; he also provided exhaustive interpretations of these experiments. Young demonstrated that the oscillations in light waves are not longitudinal as in acoustic waves but transverse, as in vibrations of liquid particles in waves on the surface of water.

After the work of Young and other scientists, the wave nature of light was assumed to be proved beyond doubt. The theory of the world ether was treated as one of the most important achievements of 19th century science, and the existence of the ether itself was regarded as firmly established.

The entry for the ether, written at the very beginning of our century for the excellent and extremely popular Russian encyclopedia by Brokhaus and Ephron, says with complete assurance that once the experiments proved the validity of the wave theory of light, '... The existence of ether as an energy carrier where there is no matter in forms that are familiar to us, became proved and the ether ceased to be a hypothesis.' Several sentences later the author regretfully remarked that 'Nevertheless, arguments against the existence of ether are still encountered even in our time.'

We thus see that the majority of physicists firmly believed that there was a medium which permeated entire space. However, this meant that Isaac Newton's 'absolute space' was not empty but filled with ether. It was then natural to try and measure the velocity of motion of the Earth relative to the ether, and hence relative to the absolute space. If this were possible, Newton's absolute space would cease to be a pure abstraction that does not manifest itself in anything, but would become a specific object of study.

Albert Michelson, whom I have already mentioned, became interested in this problem in the 1880s. He designed an excellent high-precision instrument now known as the Michelson interferometer, which was expected, according to calculation, to solve the problem.

However, how would one measure the velocity of the Earth with respect to the ether? Indeed, since by definition the ether wind blowing against the Earth flows freely through all bodies, producing no pressure at all, unlike the ordinary wind in the air, the expected displacement of the Earth relative to the ether could be determined in the following way.

Let us send light signals in a laboratory moving together with the Earth through the ether, along the direction of the motion, so that these light pulses return to the light source after being reflected by a mirror. Let us refer to them as signals A. Another set of signals B will be sent at right angles to the motion of the Earth. Signals B, reflected by another mirror at the same distance from the source as the first one, also return to the source. If the Earth is at rest relative to the ether, the signals A and B will obviously spend the same time traveling from the source to the mirror and back. If, however, the Earth is moving, then it is easy to calculate that these times will be slightly different. Signals B will need slightly less time to travel. Knowing the dimensions of the instrument and the delay time, it will be a straightforward matter to calculate the velocity of the ether wind blowing against the Earth because of its motion.

In Michelson's instrument, the path covered by the light signals was about 22 meters. If we assume that the velocity of the ether wind is the same as the velocity of the Earth on its orbit around the Sun, then the delay time of signals A was calculated to be only about three ten-thousandths of a millionth of a millionth of one second (three divided by one followed by sixteen naughts).

The instrument was so perfect and precise that it was capable of measuring a delay even a hundred times smaller!

Of course, the Earth moves in the ether not only along its orbit around the Sun but also moves with the Sun, together with the Solar System as a whole. Hence, the direction of the ether wind is not known beforehand. The experimenters were able to take that into account too. They made their instrument, which was floating in a pool of mercury, rotate slowly, changing its orientation. Finally, it could not be excluded beforehand that the orbital motion of the Earth at the moment of measurement was accidentally compensated for by the displacement of the Sun in the opposite direction. To exclude such a coincidence, experiments were repeated every three months, when the direction of the orbital motion of the Sun had changed considerably.

In 1887 Michelson and Morley published the results of a series of their most accurate measurements carried out with this instrument. They failed to detect any ether wind. At this time, Michelson wrote to the famous British physicist John Rayleigh that he had completed an experiment aimed at measuring the relative motion of the Earth to the ether, and that the result was decidedly negative. This result was baffling for everybody. Michelson was openly disappointed. Many people tried to find imperfections in his experiments or to reformulate the theories of the world ether; other experiments were conducted, including experiments on detecting the ether wind in the mountains where, according to the same hypothesis, the effect of the ether wind would be more pronounced. But it was to no avail. This great disappointment for Michelson turned out to be the greatest triumph of his life. The negative result meant that the ether not only leaves the motion of heavenly bodies unaffected (this was clear even before these experiments) but that it does not affect experiments with light either. Hence, it was an invention, a fiction!

However, the Michelson–Morley experiments were not only a death blow to the theory of ether. Their significance was much larger. In fact, these experiments proved that the motion of the Earth does not affect the velocity of light: it remains constant in all cases. Note that this conclusion was independent of the nature of light.

Nevertheless, what is light if it is not a vibration of an as yet unknown world-permeating medium, of a putative ether?

By the end of the last century, physicists were quite ready to answer these questions. The work of Michael Faraday, James Clerk Maxwell and Heinrich Rudolph Hertz proved that light comprises oscillations of the electromagnetic field, which can propagate through space as electromagnetic waves and needs no medium, no ether; it became clear, therefore, that nothing in nature can be put in correspondence with this 'ether'.

It was thus concluded that light in the form of electromagnetic waves propagates through space without the mediation of any ether.

The Michelson–Morley experiments and numerous other experiments demonstrated really surprising properties of light. It was found that regardless of whether the observer moves towards a light beam or recedes in the opposite direction, the velocity of the beam relative to this observer remains unchanged! (Note that with the advent of lasers, it was possible to confirm experimentally that the velocity of light is independent of the velocity of light sources to within 0.03 mm/s.) In Michelson's time this was quite incomprehensible. Indeed, it was quite clear that if a car is moving along the road at a speed of 60 km/h and the observer drives in another car towards the former car, then the relative velocity of approach towards the observer is 120 km/h. This is indeed so. In this example velocities simply add up. However, if one of the cars is replaced with a light beam, the answer is dramatically different. The velocity of approach to a light signal is unchanged by the observer's motion.

A well-known Polish physicist Leopold Infeld wrote that the famous Michelson–Morley experiment '... has ultimately proved that there cannot be different velocities of propagation for light... that these velocities are identical in all directions and their value is c, which is the velocity of light and which, in the most strange manner, remains itself, ever constant, ever unalterable.

This result was catastrophic for the mechanistic view.'

Indeed, this was a really crushing blow to familiar notions. It was later understood (we are going to talk about it in subsequent chapters) that Michelson's experiments in fact demonstrated an inevitable conclusion: the properties of space and time undergo changes as the velocity of motion becomes very high.

This discovery, which signified a revolution in natural sciences, was made in 1905 by Albert Einstein.

4

The pace of time can be slowed down!

Here unfolds the story of the momentous achievements of science in the 20th century. I would say that the most impressive discovery was made at the very beginning of the century by Albert Einstein when he created relativity theory. He showed that there does not exist any 'absolute time', no unified unchangeable river of time which impartially carries all events occurring in the Universe.

Academician A. Alexandrov of the Academy of Sciences of the USSR wrote: 'Einstein's greatest discovery which became the cornerstone of relativity theory and a turning point in the general physical and philosophical interpretation of space and time was the revelation that nature knows no absolute time.'

Evidently, time behaves as a river with constant, unchangeable flowrate only in the habitual conditions of relatively slow motions and not very high interaction energies. Its properties are very different under very unconventional conditions! We will discuss this later in great detail.

The discovery of the relative nature of time is contained in relativity theory that Einstein created in 1905. An enormous number of books have been written about Einstein, definitely more than about any other physicist. Several factors explain this. I will quote the opinions of several well-known scientists who knew Einstein personally, and also Einstein himself; these sources may help, to some extent, to reconstruct the image of this personality and to understand the causes of his immense popularity.

First and foremost, he was a truly great researcher, and his discoveries dealt with the most mysterious properties of time and space. The scent of mystery invariably attracts those who wish to ponder the meaning of the world and our being in the world (and who are sufficiently strong to find time for this in the perpetual bustle of life). The USSR theoretical physicist academician Igor Tamm wrote:

Einstein, whom Lenin regarded as one of the greatest
revolutionaries in natural sciences, is rightly compared with
Newton. I am of the opinion that this comparison is correct
not only in the sense that Newton's and Einstein's
discoveries signify pinnacles in human striving to
comprehend nature, and that these pinnacles tower over
300 years of history of development in sciences and directly
talk to each other. I think that Newton and Einstein can also
be compared in the sense that Newton laid the foundation
of modern natural sciences while Einstein's creation, his
relativity theory, completed the edifice of classical physics.

In Soviet times, a reference to Lenin's authority was regarded
as the highest commendation. Furthermore, I am aware that some
Soviet physicists quoted Lenin in order to shield the progress of relativity theory in our country from very vigorous attempts to declare
Einstein's creation a 'bourgeois, idealistic anti-science'; during one
period, this onslaught looked very realistic. Well-known Moscow
astrophysicist J. S. Shklovsky wrote that the '"bureaucratic war-
riors for the purity of Marxism" were admonished "from above":
bosses realized that the military potential of the country is impossible without true physics'. I should remind the reader that this was
the period when the USSR was developing its rocket and nuclear
weapons.

I will return to Einstein's discoveries later. However, the greatness of these discoveries cannot fully explain the scale of his global
fame, a fame that has not ebbed throughout the 20th century. This
last observation is especially surprising since the ever-changing
fashion of our time never ceases to generate new idols.

The decisive point was Einstein's personality. Soviet writer
V. Kaverin once remarked: 'Above all others, I value in people kindness and courage. We may agree that a combination of these features makes a man a decent human being. These two qualities must
inform his moral stance.'

I believe that these words give a pithy formulation of the concept of a 'fine man'. It is fairly difficult to withstand the test for these seemingly simple and clear criteria over the whole length of one's life. Not everybody succeeds in it, but so many do not even try.

Albert Einstein was kind and courageous. People who knew him well say that his kindness stemmed from his extraordinarily clear mind and was not subject to surges of feelings and emotions. Einstein helped numerous people. The fates of scientists who suffered persecution in Germany after Hitler came to power were especially close to his heart. The Polish physicist L. Infeld wrote in a magazine *Tworczosc*: 'Never in my life could I witness so much kindness completely devoid of emotion. Although only physics and the laws of nature lifted Einstein to true emotions, he never refused calls for help if he thought that help was really needed and concluded that this help could be efficient. He wrote thousands of recommendations, gave advice to hundreds of people, spent hours talking to a lunatic whose family wrote Einstein that he alone could help the afflicted man.'†

Is not this an outstanding example of kindness and mercy which are often in very short supply in our frequently cruel life? This purity of goals is all the more valuable because it emanated from a man who seemed to exist in the world of abstract formulas and far removed from real life. In fact, he *was* far from the little daily worries – in that area which did not touch the primary human values. He tried to spend an absolute minimum of time on the trivia of life, thus saving time for the really important. He wore his hair long to minimize visits to the barber, preferred a leather jacket to avoid shopping for a new suit as long as possible, decided to forgo socks, suspenders and pajamas. Immersed in his thoughts, he often ate

† Translator's comment: We follow the Russian translation of the original text in Polish in: *Einstein and Today's Physics* ed. E. B. Kuznetsova (Moscow: GTTL) 1956 (in Russian).

automatically, paying no attention to what he swallowed. And he *was* courageous! He never flinched from defending the just cause, never bothered whether his actions may have led to personal troubles. He took part in anti-war demonstrations even during World War I. All his life he agitated for peace and unity of people.

Being worried that Hitler's Germany could develop the atomic bomb, Einstein was one of those who helped initiate the work on this weapon in the USA.

He realized, even before the first atom bomb was exploded, the scale of the threat brought by nuclear weapons to mankind, and thus advocated international control of nuclear arsenals.

I will give here an excerpt from his letter to Infeld, written in 1950 but sounding topical and wise almost fifty years later.

> You know well that I hold the striving to true peace in the highest esteem. I believe that in the terrible situation we are now facing the direct measures that became increasingly popular have no chance of success because confidence in honest intentions of the opposing side declined everywhere. I have no immediate suggestions. Only some individual steps by the sides can be considered at present, which promise to revive the confidence without which there may be no approaches to sustaining international security.

Is it surprising, therefore, that this man excited hatred in people who were his antitheses. Such people went as far as founding an anti-Einstein organization, and called for having him murdered.

Here is how Einstein defined his moral position in a letter to his friend, the German physicist Max Born.

> What is required of a man is to show an example of purity of ethical principles and have courage for retaining these principles in a cynical society. I kept trying to live in this way for a long while – with various degrees of success.
>
> *Naturwissenschaften* **42** 425 (1955)

51

Max Born concluded: 'This is about ... the purity and honesty in thought and feeling. We bow our heads to Einstein as example and teacher in both respects.'

I will also mention Einstein's attitude to his unusual fame: he seemed to be absolutely indifferent to it. I will again quote L. Infeld:

> Einstein was utterly indifferent to his fame: he may be
> a unique person who was not affected in the least by the
> greatest imaginable glory. The Nobel Prize medal, together
> with other medals and dozens of honorary diplomas were
> kept in a box in his secretary's room, and I am quite sure
> that Einstein had no idea of what the Nobel Prize medal
> looked like.

Einstein's long-lived fame which was his fate when he lived and has kept growing since he died, finds its explanation in the complete harmony of his greatness as a scientist and his striving to defend the oppressed and help the progress of humankind. The combination of these impeccable moral standards with amazing discoveries of mysterious properties of nature produced a firm foundation for his fame. Lev Landau, Soviet theoretical physicist, winner of a Nobel prize for physics, was of the highest opinion of Einstein. This is how Vitaly L. Ginzburg remembers his words:

> Landau had a scale of merit in physics. The scale was
> logarithmic (class 2 meant achievement smaller by a factor
> of 10 than that of class 1). Among physicists of our century,
> only Einstein had class 0.5, Bohr, Dirac, Heisenberg and
> some others were class 1... As you see... Landau placed
> Einstein above all physicists of our century, and this
> opinion is simply unassailable.

The reminiscences of people who knew Einstein well and the words of outstanding physicists quoted above are all laudatory to the highest possible degree. They may be leading the reader to a picture of a perfectly ideal person, devoid of any drawbacks. Was Einstein such an ideal human being?

This is very unlikely. Being ideal is not for a real, non-fictional, living person. Such is the 'logic of real life'.

For some years now, I have begun to hear muted statements by my German colleagues that in his private life Einstein was anything but ideal. Even books based on documents have begun to appear recently, which state that Einstein did have many drawbacks typical of ordinary people. It is not easy to sort out nowadays what is true, what is rumor and gossip and what is pure invention. Myth is always created about great historical figures.

In this connection, it will be of interest to recall what Einstein himself wrote in his letter to Morice Slavin on March 28, 1949: 'Very often we can only see an outstanding personality through a haze of sheer fog'.

My own experience has taught me that reconstruction of the personal life of a famous figure person is especially difficult. It happened when together with my colleague Aleksander Sharov, I worked on a biography of Edwin Hubble (*Edwin Hubble, the Discoverer of the Big Bang Universe*, Aleksander Sharov and Igor Novikov, Cambridge University Press, 1993). I quite agree with a remark by the well-known American astronomer Alan Sandage quoted in that book: 'It seems to me that from the scientific standpoint, we know a great deal of what he did, and that was all documented in the records and his publications. There is no question about the great things he did, but his personal life will be quite a bit more difficult to reconstruct.'

I would like to end the short digression on Einstein's personality with two of his comments that he made in a letter to the Polish physicist L. Infeld, written in 1950 (see footnote to p. 50).

The first of them sounds very fresh today. 'Before our time, man was essentially a plaything in the hands of blind forces of nature;

nowadays we are a plaything in the hands of bureaucracy. Nevertheless, man accepts this role. You know Lichtenberg's aphorism: "Man learns little from experience since each new blunder appears to him in new light".'

The second passage characterizes Einstein's attitude towards life in general and brings out clearly the inherent harmony of his inner world, which was always at one with the natural run of processes dictated by the laws that rule the world. 'Life is an exciting and splendid spectacle. I love it. However, I wouldn't be greatly impressed if I found out that I was to die in three hours. I would think how to use best these three hours left for me. I would then put my papers in order and lie down to die.'

Such was the creator of relativity theory. Now, what does this theory tell us?

The theory is based on two postulates which generalize the observational data. The first of them states that uniform translational motion cannot in any way affect physical phenomena.

We have already met this statement when discussing the Galilean principle of the relativity of motion. However, Einstein's postulate brings an important generalization to it. The reader will recall that Galileo was speaking only of mechanical phenomena: the motion of objects thrown by hand, the flight of flies etc. These were not affected by the motion of the ship. However, Einstein emphasized that not only mechanical phenomena but all the others, such as electromagnetic phenomena, will proceed in the stateroom of a moving ship exactly as they do in a ship at rest.

The second postulate of relativity theory states that the speed of light in vacuum is always the same, regardless of the motion of the light source or light detector, and equals (by today's data) $c = 299\,792.458$ km/s.

We accept the first postulate as something very natural; the second one, however, meets with serious doubts.

Indeed, imagine a spotlight and an observer to be at rest relative to each other, the observer measuring the speed of light c arriving from the spotlight. It seems logical that if the observer moves towards the light beam, the speed of light relative to him must increase and be higher than c. We know, nevertheless, that numerous experiments have proved that this expectation is wrong and the speed of light remains unchanged. All the same, it will be useful to discuss the situation further.

Let an observer in a rocket moving at high speed send a light signal from ceiling to floor; after reflection from a mirror placed on the floor, light returns to the ceiling (see figure 4.1). The observer in the rocket sees that the light beam travels in both directions along the same trajectory. As for the non-moving observer outside the rocket, he records that the light beam moving with the rocket follows a V-shaped trajectory which is longer than the simple 'up and down' path for the observer in the rocket. Hence, the velocity of the light signal must seem higher for the outside observer than for the observer in the rocket.

Stop! Recall that the velocity of a signal is the ratio of the path length to the time of travel. The path is longer for the outside observer, that is true. Doesn't this mean that the velocity is also higher? This would be so if the time of passage were identical for both observers; doesn't this equality appear obvious? Indeed, in both cases this is the time of signal propagation 'there and back again'. True, of course, but only if we assume that time flows identically for both the moving observer and the one at rest. Is there any basis for doubt here? Isn't time the duration that is common for everyone and everything?

Here lies the snag. We tacitly assume that time does flow indistinguishably for all observers. What is it, however, which makes us accept this assumption?

It is our accumulated experience that does it. In all the situations

Fig. 4.1.

we have ever experienced, clocks ticked at the same pace (provided they were in good working order) regardless of motion; in other words, time flows identically. When the journey is over, both the stationary clock and the clock that moved show the same time. However, this only happens because we deal with slow motions! The Michelson–Morley experiments and other later experiments gave the first indication that it is wrong to assume that time flows at the same rate in fast motions.

Albert Einstein was the first to clearly recognize this fact. To do so was far from easy. Not only was it necessary to analyze all the results of numerous experiments; one had to achieve the most important thing: to disengage oneself from habitual stereotypes

56

of thinking, which had been building in science for so long and seemed so unshakable.

The conclusion made by Einstein's theory was as follows. If the observer studies processes in a 'laboratory' that moves at high speed relative to him, these processes unfold at a lower rate than the same processes in his stationary 'laboratory'. For example, a clock on a fast-moving rocket ticks more slowly, the astronaut's heart beats more slowly, all biochemical processes in his body are slower, electrons in atoms oscillate more slowly, etc. Absolutely all processes go at a lower rate, hence time itself has slowed down. The higher the velocity of the spaceship, the greater the time slowdown. As the rocket velocity approaches that of light, the rate at which time flows tends to zero (time stands still) and all processes become infinitely long. If the velocity is low compared to the speed of light (say, as low as our ordinary terrestrial velocities), the time slowdown is so minute that it goes absolutely unnoticed.

The reader may have a suspicion that this slowdown of processes is only apparent when an observer regards the rocket hurtling by at a high speed. At different moments of time the rocket is at different distances from the observer, and the light that carries the image of processes on the spaceship to the observer, leaves the rocket at different moments of time and covers different pathlengths to the observer, thus taking different times to travel. Could it be that light signals have different delays when they reach the observer and this warps the true picture of what happens in the spaceship?

No, all that was said about time slowdown holds true for the actual rate of processes and does take into account unequal retardation of light signals arriving at the observer. In other words, this is the true slowdown of everything happening on the rocket as recorded by the external observer.

This effect of time slowdown may be very difficult to be at ease with for anyone who hears of it for the first time. I tried to sort it out for myself when still in the fifth form, and it took years before I was able to understand it all to my satisfaction. I will return to difficulties encountered in comprehending relativity theory.

The next question is: are there any observable facts which prove that time does flow less fast on a rapidly moving body? Yes, such facts are known, and they are the weightiest arguments in favor of this conclusion of relativity theory.

I have emphasized already that time slowdown becomes appreciable only when the body moves at a velocity close to the speed of light. Enormous energies would have to be expended to accelerate large bodies to such speeds, so this is unfeasible in terrestrial conditions. Elementary particles are a different proposition. Physicists learnt to accelerate them a long time ago to nearly the speed of light in special devices called *accelerators*. The study of processes involving fast particles completely confirmed the results of relativity theory.

Here is what happens in one of the experiments with particles known as charged pi mesons. These particles are unstable and, being created in certain processes, live only a very short time and spontaneously decay. If very many such particles are born and all are moving at low velocities, one half of them decay in just seventeen billionths of a second. This is the so-called decay half-time. Seventeen billionths of a second later one half of the survivors decay, and so on.

However, if pi mesons are accelerated to a velocity of about nine-tenths of the speed of light, then time for them begins to flow more slowly and by our clocks their lifetimes increase. This is indeed observed in real experiments. The decay half-life of these fast-moving particles is found to equal thirty-nine billionths of a second, which is more than twice the decay half-life of pi mesons at rest. The result is in complete agreement with the conclusions of the theory.

Another example. Particles with very high kinetic energy are constantly arriving in our atmosphere from cosmic space. These particles are called *cosmic rays*. The interaction of cosmic rays with particles of the upper layers of the atmosphere creates a host of new elementary particles. Among them we find the so-called muons. These are also very short-lived particles. They decay after only two millionths of a second. This is their lifetime when these particles are at rest with respect to the observer. Having been created in the upper atmosphere, muons may have velocities of about 99% of the speed of light. If time for them did not slow down, they would cover only about six hundred meters during their allowed two millionths of a second. In fact, measurements show that they traverse many thousands of meters before decaying. This happens because time on such fast-moving particles flows approximately seven times more slowly and 'for us' they live so much longer, having time to cover such a long distance.

I can give an even more impressive example. Among particles in cosmic rays we find protons (nuclei of hydrogen atoms) that move so fast that their velocities differ infinitesimally from the speed of light: the difference occurs only in the twentieth (sic!) non-zero decimal after the decimal point. Time for them flows more slowly than for us by a factor of ten billion. If, by our clock, such a proton takes a hundred thousand years to cross our stellar system – the Galaxy – then by 'its own clock' the proton needs only five minutes to cover the same distance.

The reader may counter that, well, this is true for the tiniest specks of matter. But is appreciable retardation of time flow ever observed in the motion of macroscopic bodies?

Yes, such phenomena are well known. They are observed by astronomers. At the end of the 1970s, a group of American astronomers headed by Bruce Margon discovered super-fast ejections of gas jets from a binary stellar system known as SS433. The stars of this system, tied together by mutual gravitational attraction, revolve around their common center of mass. The system lies at a distance of about ten thousand light years from the Earth. (One light year is the distance travelled by light during one year; roughly, it equals ten thousand billion kilometers.) Owing to complicated processes that I will not discuss here, two powerful gas jets are emitted from the system in opposite directions at a velocity of about eighty thousand kilometers per second each. This is almost a third of the speed of light! To give you some idea of the power of the gas flows in SS433, note this figure: each second the jets throw out a billion billion tons of gas.

With the velocity being so high, time must flow in the jets several percent slower than for us. This slowdown is not as dramatic, of course, as for fast elementary particles, but it is appreciable and can be easily measured. The jets of ejected gas consist mostly of hot hydrogen. Hot hydrogen under terrestrial laboratory conditions emits electromagnetic waves of strictly defined frequency. If this emission from hydrogen is analyzed by a spectrometer, one finds that hydrogen gas emits in certain lines of certain color, which correspond to well-defined frequencies of oscillating electrons that emit light waves.

As time is slowed down in the fast jets, the frequencies of spectral lines emitted by hydrogen must decrease, and the emitted light get redder. This is indeed observed.

Note that when the source moves with respect to the observer, the frequency of light, that is, its color, changes also for a reason not directly connected with relativity theory. This is the Doppler effect that we all know from school days: as the source moves towards us, the frequency of light waves received by us is increased and the light grows more violet. If the source moves away from us, the light is reddened. There is no doubt that these effects are not connected with the slowdown of time flow.

The Doppler effect is also observed in the stellar system SS433. However, this system is so structured that the direction of jet ejection is constantly changing in space, with a period of 164 days. Twice during this interval the jets are moving exactly at right angles with our line of sight. At these moments, the gas in the jets is neither approaching nor moving away from us, and the Doppler effect causes no frequency changes. (I ignore the relatively low velocity of motion of the entire SS433 system with respect to the Solar System.) It is at these moments that astronomers observe the reddening of hydrogen spectral lines that is caused solely by time retardation owing to the fast motion.

It should also be mentioned that the slowdown due to fast motion has been measured by a highly accurate atomic clock placed on an ordinary airline jet plane. True, some other subtle effects changing the 'ticking' of the clock also had to be taken into account.

We can summarize now. However paradoxical we may regard Einstein's conclusion – that from the standpoint of an external observer (relative to whom a body moves) time on this fast-moving body is slowed down – this has been conclusively verified and confirmed by direct experiments, and is now beyond any doubt.

Time is therefore relative. Absolute time is something non-existent.

We have seen already that the speed of light plays a special role in Einstein's theory. This is the velocity at which all electromagnetic oscillations propagate through the vacuum regardless of frequency – from low-frequency radio waves to visible light, to high-energy x-rays, to ultra-hard gamma radiation. This velocity remains unchanged relative to any observer.

The theory states that the speed of light is the largest of all velocities allowed in nature. The Soviet astrophysicist A. Chernin found an excellent image for this: 'This is the absolute record of velocity'.

What is the obstacle that prevents a body from being accelerated to a velocity above the speed of light?

Let us follow what happens to a body if it is subjected to a constant force which continuously accelerates it to a greater and greater velocity. Isaac Newton assumed that if the force acts for a sufficiently long time, the body can acquire an arbitrarily high velocity. However, Einstein's theory shows that as velocity grows, so grows the mass of the body, which is a measure of inertia, that is, of the 'resistance' of the body to the force applied. This growth of mass is a consequence of Einstein's famous discovery of the equivalence of mass and energy. As velocity goes up, and hence kinetic energy increases, mass increases too. But if mass increases, the acceleration produced by the force inevitably decreases. As the velocity approaches the speed of light, the mass goes to infinity and no force can make a body overcome the barrier of the speed of light. The speed of light sets the limit for the propagation of any field and, in general, for transmission of any information.

We should now look at another property of time discovered by Einstein. Imagine a train traveling at a very high speed. One physicist is standing at the midpoint of the train on an open flatcar. The other physicist stands on the ground and the train is rushing past. Signal lights that can be turned on when required are fixed to the front and rear points of the flatcar. Let us conduct an experiment by switching on the signal lights in such a way that light from both lamps reaches the 'train physicist' simultaneously and exactly as he is passing the 'ground physicist'. Both the 'train physicist' and the 'ground physicist' see both flashes at the same moment. What conclusions will the two make about the times when the lamps were fired?

The 'train physicist' says: 'I am standing in the middle of the flatcar at equal distances from the car ends. I saw the flashes simultaneously and, since the speed of light is always the same and equals c, the lamps have obviously flashed simultaneously.'

The 'ground physicist' comes to a different conclusion: 'I saw the flashes simultaneously, when being right against the midpoint of the flatcar, with the lamps at equal distances from me. Light needs some time to reach me, the train still moving during this interval. Hence the tail lamp of the flatcar was farther from me than the front one when light left it. Consequently, light emitted from the two lamps covered unequal lengths (that from the tail lamp traveled the longer path). The speed of light is always the same and equals c. I saw the flashes simultaneously, so the signal from the tail lamp must have been emitted earlier than from the front one. The flashes were not simultaneous.'

We see: what was simultaneous on the fast-moving body, was not simultaneous for the physicist on the ground.

The seemingly simple and clear concept of simultaneity of two events is found not to be so obvious after all. There is no absolute simultaneity. This concept is relative and depends on the motion of the 'laboratory' body with respect to which we consider the events; physicists say that it depends on the frame of reference.

If events are simultaneous and take place not far from one another in space, even comparatively fast motions make them non-simultaneous by only a tiny interval of time. In our day-to-day life, therefore, simultaneity is absolute, obvious and independent of any motion. For instance, the statement that a train left the platform simultaneously with the clock on the town square showing twelve o'clock sounds identical for all practical purposes and perfectly clear to an observer parked close to the railway station platform and for another who drives through the square. The situation is very different for events that are separated by great distances and regarded with respect to observers that move fast relative to one another. For example, a statement similar to the earlier example, made by a person on the Earth – 'A supernova exploded today at noon in the Triangulum constellation in the Galaxy' – may not be true for an astronaut traveling in a fast-moving rocket.

Relativity theory has established that the notions of 'now', 'before' and 'after' have simple meaning only for events occurring in the vicinity of one another. For events that are separated by huge distances the meaning of 'before' and 'after', 'earlier' and 'later' is unambiguous only when a signal propagating at the speed of light has had enough time to travel from the place of the first event to the place of the second one. If, however, the signal is still on its way, the 'before'–'after' relationship is ambiguous and depends on the state of motion of the observer. What is 'earlier' for one observer, may be 'later' for another, moving with respect to the former. Such events cannot be causally related, nor influence one another. Otherwise an event that was the cause of another (and thus had to precede the latter event) could be regarded by some other observer as having occurred after its consequence.

Such properties of time are related in the most direct manner to the fact that the speed of light in vacuum is always the same and independent of the motion of observers, and that this is the maximum possible velocity. Nothing in nature can move faster than light in vacuum.

Finally, I shall mention one more corollary of relativity theory.

Fast-moving bodies contract in the direction of their motion, while their dimension at right angles to the motion remains unchanged. This contraction is absolutely unnoticeable at low velocities but is large at velocities near the speed of light.

These consequences of relativity theory dramatically change our notions of space and time.

A question is very likely to be prompted at this point: 'What are the feelings of an astronaut sitting in a rocket moving at such high velocities? How will he (or she) perceive the changes in time and length that are apparent to an external observer?'

The answer is obvious: the astronaut will feel nothing at all! Indeed, as far as the external observer is concerned, both the pulse rate of the astronaut and the rate at which his clock is ticking, as well as all other processes, are slowed down to the same degree. Hence, pulse rate and clock ticking are synchronized relative to each other as before. Say, his heart still beats once each second. In his own time flow (known as 'proper' time) everything proceeds as in a rocket at rest. However, the flow of 'proper' time changed the flowrate with respect to the external observer. It is thus clear that the 'time river' does not flow at a permanent rate everywhere.

The astronaut cannot discover the contraction of the longitudinal dimension of his rocket either. Indeed, any meter-long stick or any other reference with which he might wish to measure a length will shrink equally and the number of such unit lengths along the contracted rocket will be the same as before it picked up high speed.

The astronaut has thus discovered nothing at all! He does not feel that his velocity is so high. Of course, this conclusion is in complete agreement with the first postulate of relativity theory which states that everything in a fast-moving rocket happens exactly as in the rocket at rest.

Since uniform motion is relative and there is no absolute motion, the astronaut has every right to regard himself as being at rest and the observer on the Earth as flying in the opposite direction. Furthermore, the astronaut assumes that time ticks more slowly on the Earth than in his rocket. The reader for whom this is the first encounter with relativity – and who has mostly forgotten what school teachers explained about it – may have a legitimate question here: 'How can that be? The terrestrial observer concludes that the astronaut's time ticks more slowly, while the astronaut believes the opposite is true. Where lies the truth? I can accept that time can slow down, although this is hard to digest, but has it slowed down for the astronaut or for the terrestrial observer? To quote A. A. Milne's Winnie-the-Pooh, "Either a tail is there, or it isn't there. You can't make a mistake about it." There must be an unambiguous answer to this question!'

Actually, there must not, however strange this may sound. In fact, this is not difficult to explain. For comparison, recall Galileo's argument concerning the fall of bodies in the cabin of a moving ship. For a passenger in the cabin, an object released from the hand falls straight down to the feet. For the external observer the falling object moves together with the ship and its trajectory is a parabola. One could ask: 'Is the body moving along a straight line or is it tracing a parabola?' Obviously, asking what is the 'true' shape of the trajectory is meaningless. The trajectory of a body depends on what one defines it relative to. It is 'truly' straight for a person in the cabin and 'truly' parabolic for the external observer. There is no contradiction here.

The same holds for time slowdown. Astronaut's time flows 'truly' slower for an observer on the Earth, while all events on the Earth are 'truly' slower for the astronaut. Again, there is no contradiction here. This all follows from relativity theory.

Of course, this is not very easy to digest. However, Einstein's theory is an inescapable corollary of experimental observations. In such situations it is useful to recall one of Sherlock Holmes' mottoes ' When you have eliminated the impossible, whatever remains, however improbable, must be the truth' (Conan Doyle *The Sign of Four*).

Those readers who have failed to achieve complete clarity immediately, in understanding all this, need not despair. After Einstein's discovery, even quite a few very prominent scientists took a long while to come to terms with his theory. As for 'average' scientists, to say nothing of people not familiar with physics, they faced enormous difficulties in accepting the ideas that virtually overturned habitual notions. Many tried to uncover errors and contradictions in the theory.

Attempts of this sort had not ceased even decades later. For instance, in 1931, a quarter of a century after Einstein's theory was published, a book was published in Leipzig, entitled *100 Authors Against Einstein*. One hundred expert authors of the book completely rejected relativity theory and its corollaries. The legend has it that when Einstein was told about the book, he smiled and, as always phlegmatic in such situations, remarked that if his results were wrong, the arguments of one expert would be amply sufficient. (I retell this anecdote after the description in *Introduction to Relativity Theory and its Applications to New Technologies* by N. I. Goldenblat and S. V. Ulyanov (Nauka, Moscow, 1975).)

Of course, there are no contradictions in Einstein's conclusions. For serious scientists, arguments against relativity theory have long become pieces of past history. The theory lies at the foundation of all modern physics. It is used to design gigantic accelerators of elementary particles and atomic power stations; it was tested in such monstrous experiments as explosions of nuclear bombs.

It should be mentioned that school and college students of today usually have little difficulty in mastering Einstein's theory; they achieve this with greater ease than physicists of the beginning of the century and even people of my generation who were born closer to mid-century. The reason for this is quite clear: the very style of scientific reasoning has greatly changed as we have moved towards the 21st century.

I have mentioned already that in times when new seminal ideas are about to break through in science, it is typical for several scientists to come very close to formulating the emerging relationships and to interpret some of their properties. However, someone of real

67

genius then comes up with the ultimate formulation of the new understanding. This was the fate of relativity theory as well. Some formulas of its mathematical equipment had already been written in the last century, by the end of the 1880s. The Dutch physicist Hendrik Lorentz and the French mathematician Henri Poincaré came very close to creating the theory. But the hardest step that demanded maximum courage, one that revolutionized the notion of time and space, was made by Albert Einstein. In 1912 Hendrik Lorentz was reminiscing about his attempts even before 1905 (that was the year in which Einstein's paper was published) to resolve contradictions that followed from experimental results. He wrote that in his paper written in 1904 he failed to derive in a complete and satisfactory manner the transformation formulas of Einstein's relativity and that this led to the weak and helpless arguments one finds in his paper. Lorentz added that Einstein's great achievement was that he produced the first formulation of the relativity principle as an all-encompassing, strict and accurately functioning law.

Another remark is in order here. Beginning in 1990, some authors have tried to find a foundation for the rumor that Einstein's first wife Mileva Marić played an essential part in the creation of special relativity. I do not think this story has any credibility at all. I will quote the opinion of an expert in the history of science, Harvard University professor Gerald Holton (*Physics Today*, August and September, 1994):

> Careful analysis by established historians of physics, including John Stachel, Jürgen Renn, Robert Schulman and Abraham Pais, has shown that scientific collaboration between Mileva and Albert was indeed minimal and one-sided.

The lively discussion of this topic that flared up at the beginning of the 1990s was most probably caused by the thirst of a fraction of the reading public for science-history sensations.

5

Time machine

Which of us was not immersed in our youth in Herbert Wells' famous short novel *The Time Machine*? The protagonist of this story uses a device that can travel in time to visit a very remote future of the Earth. Wells also imparted to this device the property of reverse motion into the past.

A large number of books have been written which fantasized about the possibility of freely visiting the past and the future. In all likelihood, their authors were never in doubt that their inventions belonged to pure imagination and treated this as nothing more than a literary stratagem.

The entire experience of mankind and scientific knowledge have made inevitable the conclusion that travel in time is impossible. Space is where motion is allowed. Say, travel on the Earth is possible in different directions and one can also return to the starting point. On the contrary, we are seemingly unable to choose the direction of motion in time, we are bound to 'float' passively in this flow. It was assumed that here lies the dramatic difference between time and space.

Einstein's discovery of the surprising properties of time in 1905 demonstrated the fallacy of the view that we are 'captives' of the river of time and thus cannot 'steer' on it; it was seen as a fruit of not knowing, as a consequence of the limited possibilities that mankind had during its preceding history. But does this mean that we are free to roam in time?

Yes and no! Einstein's theory has solved, so to speak, only half of this problem. It was shown that we can only propel ourselves 'downstream': move towards the future, leaving the flow itself behind. However, the theory revealed no 'upstream' way, no access to the past. Still, how could one reach the future, thus overtaking time?

To achieve this, Wells' personage jumped into the time machine, pressed a lever, the machine began to shake and then transferred

70

itself to different epochs, disappearing from the 'now' together with its driver.

The theory of relativity proved that this sort of traveling in time is forbidden. You would have to move through space in order to move in time. To reach the future of the planet, one has to get into the photon rocket mentioned above, accelerate it to a speed very near the velocity of light, travel through space at this tremendous speed for some time (say, a year) and then return to the Earth. From the point of view of people who were left behind on the planet, the time on the fast-moving rocket advances more slowly than their planet time. Hence, when the crew of the rocket lands at the homeport, the time lived through on the Earth is longer than that by the clocks of the astronauts; consequently, the travelers arrive in the future of their planet.

The French physicist Pierre Langevin discussed the following thought experiment in 1911. Imagine a twin brother departing on the rocket for a space voyage, leaving his twin brother on the Earth. When he comes back, he is younger than his brother who waited for him at home. For the astronaut, this is a tangible result of the voyage into the future of the Earth.

In fact, some theorists doubted that this effect was possible. They argued that Einstein's theory states the relativity of motion. Hence, the astronaut can regard himself as stationary and the Earth-bound people as speeding away in the opposite direction at the same speed. From his point of view, then, the clock on the Earth is ticking more slowly than that on the spaceship. He concludes, therefore, that the twin brother on the planet will be the younger one upon return of the rocket.

This produces an apparent paradox. Each brother considers the other as the younger one at the end of the experiment. Which argument is correct? Indeed, coming together, the brothers will identify the younger one immediately by his looks. This is the origin of the famous 'twin paradox'.

Specialists were able to sort out the situation quite quickly, so truth was out, but for the uninitiated the 'twin paradox' rumors manifested the failure of relativity theory for many years. Alas, 'reasoning' of this sort is sometimes found in literature even today. So who is older and why?

The important point is that the argument about the rate of advance of a clock is valid only from the standpoint of a 'laboratory' or, in general, bodies moving by inertia. Physicists say that Einstein's formulas (in the form he has written them) hold only in 'inertial frames of reference'. A passenger does not notice the motion only when a ship or a rocket moves without accelerating or decelerating. There is no doubt that the astronaut feels the acceleration when the rocket, say, blasts away. Hardly anybody today is ignorant of g-loads that astronauts undergo during the launch and landing stages of spaceships.

It is thus clear that the positions of a person on the Earth and an astronaut in the rocket are not equivalent. The Earth can be regarded, as a good approximation, as an almost inertial reference frame. For the space traveler to return to the home planet after a long and lengthy journey, however, it is necessary to decelerate and stop the ship, then accelerate it towards the Earth and again slow it down to land safely. Of course, the motion is not inertial during the acceleration and deceleration stages, and the astronauts undergo the corresponding loads. During these intervals of motion, the formulas written for inertial system are not applicable to the 'laboratory'-ship and the astronaut has no basis for considering the terrestrial clock to be slower.

Here I will not go into details of this process. Theorists know how to calculate time in a 'laboratory' even if it moves with acceleration. I will give the final conclusion of a physicist. There is no contradiction, and the conclusion of the observer on the Earth was correct, since his frame of reference is always inertial (with sufficient accuracy) while the rocket moved with acceleration. The 'naive' inference of the astronaut during these periods that the clock on the Earth is the slower one is wrong. Hence, the space voyager travels to the future when he returns to the Earth. The faster the motion of the rocket and the longer the flight, the more remote the future to which he is transferred.

This possibility of visiting the future is quite awesome to anyone who learns about it for the first time, while reading up on relativity theory.

When I was a third-year astronomy student at Moscow University, I accidentally noticed the 'twin paradox' among the topics offered for the term research project. Later I learnt that the adviser for this topic was a well known Soviet cosmologist A. Zelmanov.

At that time, relativity had not yet percolated into school courses of physics. Nevertheless, I had by that time already read several popular science booklets on this theory and thought that I had some notion of this paradox. True, I had no detailed knowledge of the theory itself; I remembered its 'ominous' reputation as something super-complicated and doubted that I could actually calculate anything myself. Still, the aura of mystery led me to Zelmanov.

He was a soft-spoken and sensitive man, of vast knowledge and a manner of working in a style that was rather typical for the 'old school' of the end of the 19th century. What I mean is an unhurried, thoughtful, pedantic way, when ideas are thought over for a very long time, all the calculations are extremely thorough and repeated many times, and papers are prepared for publication for years! This is so unlike the high-speed high-pressure style of today's science (in keeping with the entire life around us).

By that time Zelmanov had already suffered from the unlimited voluntaristic rule of bosses who, even though utterly incompetent in anything connected with science, ruled over it and dictated its fate. They decreed that cosmology – the science studying the structure of the entire Universe and, among other aspects, its expansion (I will talk about it later in the book) – was a pseudo-science, or non-science, contradicting or denying the 'dialectic Marxism'. At the beginning of the fifties, Zelmanov was fired from his post at the P. K. Shternberg Astronomical Institute in Moscow. When I met him, the situation had already improved and he was allowed to return to the institute.

During my first appointment with Zelmanov, he explained in detail what he expected me to calculate, what is the rate of advance for the clock on the Earth as conjectured by the astronaut, what sort of Universe will he observe from the cockpit of his spaceship etc. I could not fathom too much at that first meeting, so I started to work with the famous textbook of theoretical physics by Lev Landau and Eugenii Lifshitz: Zelmanov recommended it to me as a good preparation for tackling the problem.

A couple of weeks later I had the impression that the required chapters were pretty well understood, so I went to see Zelmanov again. He heard me out and said: 'That's just fine. You can start your calculations now.' That was a blow: 'Start calculating'. How? I did not have the slightest idea of what the first step could be. However, my adviser was a very wise tutor. He identified my obstacles immediately and hinted in a few words what to do as a start for calculating the effect involved in the motion of the reference frame 'spaceship'. I started to calculate.

Somewhat later Zelmanov advised me to tackle a fairly complicated monograph *Theory of Space, Time and Gravitation* by V. Fock. Now, quite a few things cleared up and the work went on faster; I was even able to finish the calculations in time. This was my first work in theoretical physics and it even got published several years later. It has mostly been of methodological interest but also contained original results.

Now about the results themselves. My first question was how the space voyager is going to see the Universe through the windows of his ship – this 'laboratory' hurtling through space and time.

The astronaut will observe two effects. The first of them is the already familiar Doppler effect which makes light shift to blue when we move towards the light source, and makes it 'redden' when we move away from it.

This is not all, though. The direction in which we see remote stars changes as well if the observer's speed is very high. What makes them move? Let us recall traveling in a train or a car in the rain. While we are stationary, rain drops leave vertical traces on the windows. Once we are in motion, drops leave inclined traces which tend to tilt closer to the direction of motion of the vehicle.

The picture is similar for light. For the moving observer, light rays become tilted towards his line of flight. Hence, the astronaut should see stars as if crowding towards the point to which the rocket is directed. This phenomenon is known as aberration of light; the shifts in the visible position of stars on the sky will, of course, be very large for spaceship velocities close to that of light.

I have calculated what the sky should look like from a ship traveling at a speed of 250 000 km/s. Figure 5.1 shows what sort of pattern the crew will see. For these observers, the stars in the sky rush as if to the rocket's destination point. The density of stars here will be much higher than towards the tail, where almost no stars will be seen.

The color of stars will also change as described by the Doppler effect. In the direction of motion, the passengers will see bluish stars of enhanced brightness. In the opposite direction, there will be only infrequent dim reddish dots.

How about the schedule of the voyage? In the case study that

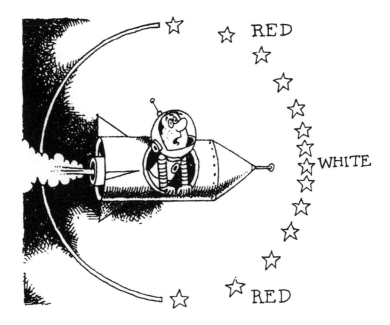

Fig. 5.1.

I chose at that time, the astronauts travel to the star which is the nearest neighbor of our Sun: Proxima Centauri lying at a distance of about 40 thousand billion kilometers (4.3 light-years) from us. By my scenario, the rocket accelerates during the first 4.5 months of the flight. The rocket engines are assumed to produce a thrust that weighs down the astronauts with a $3g$ load – three times that on the Earth. By the end of the acceleration stage, the spaceship moves at a speed of 250 000 km/s. The engines are switched off here and the ship keeps going by inertia; now the crew can contemplate the unusual view of the starry skies as described above.

On approach to Proxima Centauri, deceleration motors are switched on and the ship slows down, ultimately stopping. Then it picks up speed again towards the Sun, decelerating on the approach to the Earth. By the clock ticking on the Earth, the flight lasts about twelve years, while the clock on board the ship reports only about seven years. Having returned to the Earth, the voyagers are thrown into the future of the Earth by five years! This is how the 'cosmic time machine' works.

It is thus clear that even at very high speeds and after relatively long journeys through the cosmos, the time jump is not very large. Nevertheless, the jump is there (rather, it will be inevitable in future interstellar travel). In principle, the time jump takes place in any motion through space, even at low speeds. However, it is normally absolutely negligible. For example, when the crew of the Soviet space station *Salyut* landed in 1988 after travelling on an orbit around the globe for a year at a speed of eight kilometers per second, they stepped into the future by a mere one hundredth of a second.

In future interstellar flights, photon-driven rockets could accelerate to speeds very close to the speed of light, much closer than in the above example with Proxima Centauri, where the speed was about 80% of the speed of light. At such truly great speeds, transfer to the future may be quite serious. Imagine, for example, that astronauts set out on a photon rocket to the center of our home stellar system, the Galaxy (this will be a journey both through space and through time). For the first half of the forward leg the rocket speeds up at a constant acceleration, so that the astronauts are under constant load, twice that on the Earth, while on the second half of the leg the rocket decelerates, again at the same constant load for the astronauts. Then everything is repeated on the return stage of the voyage towards the Earth. On the whole, the return trip should take about sixty thousand years by the terrestrial clock, with numerous generations replacing one another; on the rocket, however, the crew will register only forty years! This duration is definitely within the active span of a human life, so that the people who come back to the home port may even be the same astronauts who left it. They will, however, find themselves in a very distant future of the globe. What will they find? Only science fiction writers know that. A host of problems that will arise would be social and psychological rather than scientific, and we cannot really say anything profound about them. The Polish science fiction author Stanislav Lem has described in the novel *Return from the Stars* very vividly the experiences of men who were ejected into time epochs that were very far from the familiar surroundings in which they grew and matured.

I also have to point out another specific feature of interstellar travel. At first glance, one tends to consider mankind as somehow captive in space. It may seem that an individual cannot get too far away from the spot where he or she was born, being as if 'tied' to this point in space by an invisible time chain. Indeed, nothing can move at a velocity exceeding that of light. Hence, one cannot escape by more than, say, a hundred light years over a lifetime of a hundred years. This distance stretches only to the stars nearest to our Sun.

In fact, this naive evaluation is based on a serious error: it ignores the slowdown of time for the space traveler. If this slowdown is taken into account, the ship can go very far indeed and visit very distant corners of the Universe.

The prospects are definitely exciting, aren't they?

However, inventive thinking strives for even more breathtaking horizons. Is it really necessary to break through space, making very long, very demanding interstellar trips? Could a bypass of some sort be devised?

I beg readers' patience, I will discuss this aspect later. I will only mention here a remark made after I presented a talk about the possibilities of such bypasses at a colloquium at the Institute of Theoretical and Experimental Physics in Moscow. The remark was made by Professor Lev Okun, a theorist of the highest international reputation, who said to me: 'You see, I had a walk on a starry night many years ago with one of our most renowned physicists. Looking at the stars, I mused that there simply must exist some way of reaching these stars in addition to the trivial endless flight through space. My companion looked at me skeptically and dropped: "Cut out this insubstantial fantasizing, this is a matter for fairy tales only". Isn't it wonderful that these possibilities may be opening to us now? Only in theory, of course, but opening nevertheless?' He was very glad.

I want to add to these words that science has compelled us to treat very seriously even the most extravagant theoretical predictions. Among examples of realizations of the wildest dreams are the liberation of atomic energy, space flights and many others. Things that theoreticians were scribbling just yesterday on pieces of paper, become reality tomorrow. Let us be very attentive, therefore, to physicists' predictions, even if they sound too far-fetched.

Let us be content for the time being with these remarks. I will take up the search for other paths to the stars later in the book.

I will make one more remark. Relativity theory revealed a method of traveling into the future. But what about the past? Can we return to moments gone by? Can we visit distant epochs in the history of the Earth?

I have mentioned already that the theory itself did not offer any method of achieving it. How about other theories developed after Einstein's relativity: do they promise anything?

Again I beg for patience. For the moment, I will change the question: can the past be seen? A Soviet physicist A. Chernin, well-known popularizer of science, has given the following answer: 'If we can see anything at all, it is the past.' This is unexpected, and sounds like nonsense.

Actually, the situation is quite simple. We see the surrounding world via light rays. Light needs a certain time to travel to our eyes from the object we are looking at. We thus see an object as it was at the moment when light 'left' it. Of course, the speed of light being tremendously high and the familiar objects of our daily routines being not far away, light takes negligible time to cover the distance. Nevertheless, we see objects at any given moment as they were an instant before, when light started its trip. Hence, we see them in the past! Not very distant past but past all the same.

The situation is very different when we look at objects in the sky. Light takes eight minutes to reach us from the Sun. Light from stars takes years, and light from distant stellar systems takes millions and even billions of years. We observe these systems in their very distant past. This is a duration sufficient for many stars to be born, to live out their entire lives, and for entire stellar systems to arise and evolve! We observe heavenly bodies that lie at different distance at different times in the past: the farther an object, the longer its light takes to reach us, the more distant the past in which we observe it today.

This is a very worrisome property for astronomers since stellar systems at unequal distances from us are being observed at different stages of evolution of the Universe, so comparing them involves taking into account their evolution over long stretches of time. This is not easy to do, however, since our knowledge of the laws of stellar evolution is often insufficient and all sorts of surprises are possible.

Here I leave the technical difficulties to experts and return to the topic of this book.

6

Time, space and gravitation

Everyone knows that the space of the Universe is three-dimensional. This means that space is characterized by length, width and height. The same is true for any body. Somewhat differently, the position of a point in space is characterized by three numbers known as coordinates. If we draw straight lines or planes or complicated curves through space, their properties are described by the laws of geometry. These laws have been known to man since ancient times and were compiled by Euclid in the 3rd century BC. Euclidean geometry is studied in schools as a harmonious system of axioms and theorems that describe all properties of lines, surfaces and solids.

If we wish to study not only the spatial position but also processes occurring in three-dimensional space, we need to add time as well. An event taking place at some point is characterized by the position of this point, that is, by indicating three numbers, and by a fourth number, that is, the moment of time at which the event occurred. For the event the time is its fourth coordinate. In this sense we say that our world is four-dimensional.

All this is well known, of course. Then why wasn't this formulation of four-dimensionality treated as serious and fraught with new knowledge before the theory of relativity was born? The catch lay in the fact that the properties of space and time seemed to be too dissimilar. When we speak of space, we have a static mental picture in which bodies or geometric figures are fixed at a given moment. In contrast to this, time flows incessantly (always from the past towards the future) and bodies change their positions.

Space is three-dimensional but time is one-dimensional. In fact, time was compared to a straight line even by the ancient philosophers, but this always seemed to be no more than a useful visual image without any profound meaning. Things changed drastically after relativity theory was discovered.

In 1908, the German mathematician Hermann Minkowski, developing further the ideas of this theory, said: 'From now on, space as such and time as such must turn into fictions and only some form of combining them together will retain independence.' What did Minkowski mean in this forthright and categorical declaration? He wished to emphasize two aspects. Firstly, that time intervals and spatial lengths are relative, depending on the choice of the reference frame. Secondly – and this was the more important part of his words – that space and time are connected inseparably. In fact, they are two facets of a unified entity: four-dimensional spacetime. The pre-Einstein physics knew nothing of these close ties. What are their manifestations?

The most important one is that spatial intervals can be determined by measuring the time required for light or for any electromagnetic waves to travel the distance we wish to know. This is the method used in now familiar radar. The essential point is that the velocity of propagation of any electromagnetic waves is completely independent of the motion of the source or that of the body reflecting the waves, and always equals c. Hence, the distance is found simply by multiplying the constant velocity c by the time of travel of the electromagnetic signal. It was not known before the arrival of Einstein's theory that the velocity of light is constant and thus it was expected that this procedure would be wrong.

Of course, one can choose the opposite approach, that is, measure time by a light signal covering a known distance. For example, if we make a light signal shuttle between two mirrors spaced by three meters, each jump would last one one-hundred-millionth of a second. The number of times that this unusual light pendulum has swung between the two mirrors is the number of one-hundred-millionths of a second that has elapsed.

These examples illustrate the relationship between time and space. Their respective intervals differ only in a constant, familiar multiplier 'c'.

Another, at least as important manifestation of the unity of space and time is that as the velocity of a body increases with time, the rate of advance of time decreases for the body in exact correspondence with the reduction of its longitudinal dimension (along the direction of motion). Because of this exact correspondence of these two quantities – the distance in space between two events (e.g. flashes of two light bulbs) and the time interval separating them, a simple calculation yields the quantity that is constant for all observers, regardless of the velocity at which they move, and that is independent of the velocity of any two 'laboratories'. This quantity plays the role of distance in four-dimensional space-time. The spacetime is precisely the 'unification' of space and time announced by Minkowski.

It may not be too hard to comprehend this formal unification of space and time. Imagining the four-dimensional world is far more difficult. The difficulty is not surprising. When we draw geometric figures in a plane, we usually encounter no difficulties in projecting what we want; these figures are two-dimensional (only have a length and a width).

Quite a few people have a hard time imagining three-dimensional forms in space – pyramids, cones, planes intersecting them etc. As for creating an image of four-dimensional forms, it is a very demanding task even for experts who work with relativity theory all the time.

I will quote the very famous British physics theoretician Stephen Hawking, an expert of incomparable standing in relativity theory. He says in his famous book *A Brief History of Time*: 'I personally find it hard enough to visualize three-dimensional space!' Which shows that the reader defeated by imagining four-dimensional world need not be unhappy. Experts use the spacetime concept quite successfully. For instance, the motion of a body can be shown by a line in spacetime.

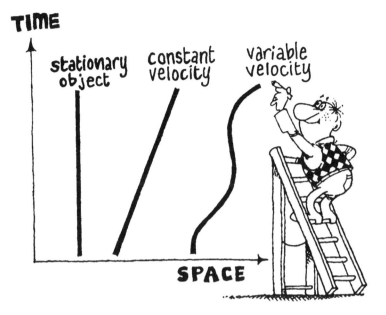

Fig. 6.1.

In figure 6.1 the distance in space in one direction is plotted along the horizontal axis and time is plotted along the vertical axis. We can mark the position of a body at each moment of time. If the body is at rest in our 'laboratory', that is, if its position does not change, our plot shows this by a vertical line. If the body moves at a constant velocity, we obtain a tilted straight line. An arbitrary motion produces a curved line, known as the *world line*. In the general case, one has to imagine that a body can move in the other two directions, not only along one axis. Its world line will picture the existence of the body in four-dimensional space.

This figure is an attempt to show that space and time enjoy identical status. The values they assume are merely marked on different axes. Nevertheless, there exists an essential difference between

85

space and time: we can stop in space, but not in time. The world line of the body is drawn vertically in our figure: as if a body is pulled along by time flow, even if it is at rest in space. This is true for all objects in the Universe: their world lines cannot stop, cannot be cut at some moment of time, since time never freezes. As long as a body exists, its world line stretches on.

As we see, there is nothing mystical in physicists' concept of four-dimensional spacetime. Albert Einstein once remarked that a non-mathematician is often given to mystical trepidation when hearing 'four-dimensional' mentioned – a feeling not unlike that produced by a ghost in a theater... while in fact no phrase is more banal than saying that the world around us is a four-dimensional spacetime continuity.

There can be no doubt that a new concept takes time to become habitual. Nevertheless, physics theorists use the concept of the four-dimensional world as their daily tool, manipulating the world lines of bodies, calculating their lengths, finding their intersection points etc. In this four-dimensional world, they develop four-dimensional geometry which is similar to Euclidean geometry. To honor Hermann Minkowski, the four-dimensional world is known as Minkowski spacetime.

Having created relativity theory in 1905, Albert Einstein worked very hard for ten more years trying to connect his theory with Newton's universal law of gravitation.

The law of gravitation as formulated by Isaac Newton is incompatible with relativity. Indeed, Newton's statement declares that the force with which a body attracts another body is inversely proportional to the squared distance between them. Therefore, if the attracting body is displaced, the distance between the bodies changes and this will alter instantaneously the attractive force acting on the other body. Therefore, Newton's gravitational force propagates through space at infinite speed. Relativity states, however, that this simply cannot be. The speed at which any force, any effect can be transferred cannot exceed the velocity of light, so that gravitation cannot act instantaneously!

In 1915 Einstein completed the development of a new theory which joined together relativity and gravitation. He called it general relativity. Subsequently, the theory developed in 1905, which could not deal with gravitation, was referred to as special relativity.

The mathematical tools of the new theory proved very complex and unorthodox for physicists of the time; as a result, it was not immediately understood nor accepted by a large number of theorists.

Despite the complexity of the mathematics involved, the basic ideas are simple (as everything that is truly important), although they were extraordinarily brave and changed the concept of space and time in an even more drastic way than special relativity did.

Isaac Newton perfectly understood that he was only able to describe the law of gravitation but that he failed to comprehend specifically how gravitation propagates from body to body, what was, so to speak, the 'mechanism' of functioning of gravitation. Newton wrote: 'But hitherto I have not been able to discover the cause of those properties of gravity from phenomena, and I frame no hypotheses (*Hypotheses non fingo*); for whatever is not deduced from the phenomena is to be called an hypothesis.' It was sufficient for him at that stage that gravitation does exist and acts in accordance with the laws as he formulated them, and that they provide adequate explanation for all motions both of heavenly bodies and of the sea.

Einstein's general relativity does reveal the 'mechanism' of gravitation. It states that gravitation is dramatically different from all other forces of nature. To clarify this point, let us resort to the following analogy. A sphere rolling on a flat surface moves along a straight line, which is the shortest line connecting any two points.

If the sphere is made to roll on a curved surface, it has to follow a curvilinear trajectory, because it is impossible to place a straight line on a curved surface. For instance, if a ball is rolling on the surface of the Earth (we assume its surface to be absolutely smooth, without mountains, valleys or obstacles), it follows the shortest trajectory on the sphere (such lines drawn on any curved surface are known as *geodesics*).

Einstein's theory of gravitation states that gravitating bodies geometrically distort the spacetime around them. I have already mentioned the difficulties in imagining four-dimensional spacetime, but if it is also curved.... However, mathematicians and physicists can live quite well without visualizable concepts. For them, curvature of spacetime constitutes a change in the geometric properties of figures and solids. For example, the ratio of the circumference of a circle to its diameter on a plane is π but this is not so on a curved surface or in a 'curved' space. The geometrical relations in them differ from Euclid's geometry. An expert can operate in such extraordinary space once he knows the laws of the 'curved' geometry.

The discovery that three-dimensional space may be curved was made theoretically at the beginning of the 19th century by the Russian mathematician Nikolai Lobachevsky and at the same time by the Hungarian mathematician Janos Bolyai. At mid-century, a German mathematician working in geometry, Georg Riemann, introduced into mathematics 'curved' spaces with four and even any number of dimensions. From that time on, the geometry of curved space has been known as *non-Euclidean geometry*. The discoverers of non-Euclidean geometries did not know under what specific conditions their geometries might manifest themselves, although some guesses were suggested. The mathematics apparatus that they and their followers developed was later used to formulate general relativity.

Einstein's fundamental idea, therefore, was that gravitating masses curve the surrounding spacetime. Let us now consider other bodies with very small masses (physicists refer to them as 'probes') which move in this curved spacetime. As before, they move along geodesics. In the non-curved spacetime geodesics are straight lines, but in a curved spacetime they are curvilinear. It is this motion along curved trajectories that we interpret as the motion caused by gravitational forces. The explanation of the gravitational field is thus the 'curved' geometry of spacetime.

Eminent American physicists Charles Misner, Kip Thorne and John Wheeler chose to begin their massive monograph (*Gravitation* 1973 (San Francisco: Freeman), 1279 large-size pages) with the following amusing story.

> Once upon a time a student lay in a garden under an apple tree reflecting on the difference between Einstein's and Newton's views about gravity. He was startled by the fall of an apple nearby. As he looked at the apple, he noticed ants beginning to run along its surface. His curiosity aroused, he thought to investigate the principles of navigation followed by an ant...
>
> His eyes fell on two ants starting off from a common point P in slightly different directions. Their routes happened to carry them through the region of the dimple at the top of the apple, one on each side of it. Each ant conscientiously pursued his geodesic. Each went as straight on his strip of appleskin as he possibly could. Yet because of the curvature of the dimple itself, the two tracks not only crossed but emerged in very different directions.
>
> "What happier illustration of Einstein's geometric theory of gravity could one possibly ask?" murmured the student. "The ants move as if they were attracted by the apple stem... Now I understand better what this book means."

The authors concluded:

> Space acts on matter, telling it how to move. In turn, matter reacts back on space, telling it how to curve.

Everything is extremely unusual in this story: a curved four-dimensional spacetime that cannot be visualized, the interpretation of the force of gravitation in terms of geometric factors. For the first time, physics is directly linked to geometry. Looking closely at physics' successes, we notice that as we come closer to our time, its discoveries become less and less conventional while its notions become less and less amenable to visualization. Well, there is nothing to be done about it: nature is extremely complex and we have to accept that the deeper our penetration into the realm of its secrets, greater and greater is the effort required for the process, including the efforts of our imagination. The word 'accept' may not be the right one here; one would rather like to emphasize that the going is getting more and more exciting even if harder and harder.

The reader will profit from information on two other facts from Einstein's gravitation theory.

In Newton's theory, the field of gravitation is determined exclusively by the mass of the body creating the field. According to Einstein's theory, all types of energy take part in creating gravitation, including energy connected with pressure and tension of the body, and the electromagnetic field. The second important prediction of the theory is that if the gravitating masses move with an acceleration, they must emit gravitational waves: we know that accelerated electric charges emit electromagnetic waves. (It is rather unfortunate that I have no chance of going into details of what gravitational waves are.)

Both these predictions of Einstein's theory, which immediately distinguish it from Newton's theory, manifest themselves only under very exotic conditions, while under ordinary conditions the effects stemming from these predictions are extremely weak and utterly undetectable. In a conventional environment Einstein's theory is practically indistinguishable from Newton's theory. On the other hand, Einstein's theory leads to conclusions completely different from anything implied by Newton's theory in very strong gravitational fields or in fields that rapidly vary in time. This will be the subject of later discussion.

Immediately after formulating his theory, Einstein pointed out three effects which, although very minute under usual circumstances, can nevertheless be put to the test in astronomical observations and used to confirm or disprove the new theory.

The first two effects involve small deviations from the calculations of Newtonian physics in the motion of planets orbiting the Sun and in the trajectories of light passing very close to it. A comparison with the observational data did reveal these effects and completely confirmed the conclusions of the new theory. By the way, the observation of Einstein's effects demonstrated that the space in the vicinity of the Sun is indeed slightly curved and its geometry somewhat deviates from Euclidean geometry.

The third effect deals with time and therefore I will go into more details here.

Einstein's theory predicts that time flows more slowly in a strong gravitational field than outside it. This means, for example, that on the surface of the Sun any clock runs more slowly than on the Earth, since gravitational pull is much stronger on the Sun. For the same reason, a clock lifted high above the surface of the Earth ticks slightly faster than a clock on the surface itself.

A considerable number of experiments were conducted to detect and quantify this exciting effect; I will describe some of them. Let us start with the observations of slowdown on the Sun.

The objects that served as 'clocks' were atoms of chemical elements. Absorption lines in the solar spectrum due to these atoms correspond to certain frequencies of oscillation of electrons, when these electrons jump from one atomic energy level to another. If time on the Sun does flow at a slower pace, the frequencies of these oscillations must decrease and therefore, the lines in the spectrum must shift towards the red end. The shift is extremely small, since the relative slowdown of time on the Sun is by only one part per two million. Hence, the frequency of a spectral line should shift towards the red end of the spectrum by the same fraction. This effect is known as the *gravitational red shift*. The experiment was to measure just this tiny shift. Astronomers would be able to measure the gravitational red shift reliably were it not for the complicating effects caused by the motion of masses of gas on the solar surface.

Unfortunately, turbulent motion of solar gas masks the gravitational effect owing to the Doppler effect, so that astronomers faced serious difficulties. The first attempts, made immediately after the prediction was formulated, were rather unsuccessful; only relatively recently, during recent decades, has analysis of the solar spectrum yielded complete confirmation of the theory. Even though the difference between the rates of time flow on the Earth and the Sun is negligibly small, the difference between the number of years that have elapsed on these two bodies is quite considerable. Both are known to have existed for about five billion years, but the Earth has clocked ten thousand years more than the Sun...

In 1968 the American physicist Irvin Shapiro measured the retardation of time flow on the surface of the Sun by a very ingenious method. He was conducting radar measurements of Mercury when this planet was on the part of its orbit around the Sun which is diametrically opposite the Earth. The radar beam going towards Mercury and the reflected signal had to pass close to the Sun, and thus took slightly longer to cover the distance than it did when Mercury was not hiding behind the Sun. This time delay (about one ten thousandth of a second) was indeed reliably measured.

Astronomers know stars which are much denser than the Sun, so that the gravity field at their surface is very much stronger: these are neutron stars and white dwarfs. Observations of the time retardation effect for the light emitted by them also confirmed the theory. Note that on the surface of neutron stars, time flows twice as slowly as on the Sun!

It is especially impressive that the slowdown in the flow of time in the gravitational field has been measured on the Earth, in laboratory conditions. This was achieved in 1960 by the American physicists Robert Pound and Glen Rebka. They compared the rate of time at the base of a tower and at a height of 22.6 m, where the clock was expected to run slightly faster. The 'clock' was in fact a set of extremely accurate instruments using the phenomenon of emission of gamma rays of precisely known frequency under certain conditions. The theory predicted a fantastically small difference between the clock rates at two heights: three ten-thousandths of one billionth of a second. Nevertheless, the difference was measured and confirmed the theory!

Sixteen years later, similar experiments were repeated but under very different conditions. In one of them, an instrument emitting radiation at a prescribed frequency (known as the hydrogen frequency standard) was launched by a rocket to a height of about ten thousand kilometers. At such an altitude, time runs faster than on the Earth's surface again by the minutest amount but the difference between rates is nevertheless one hundred thousand times greater than in the Pound and Rebka experiment. The experiment (the rocket flight) lasted two hours. However, the flight was preceded by five years of intensive work. Einstein's formula was shown to hold to within two hundredths of one percent!

At about the same time, direct experiments were carried out with clocks, or rather with super-accurate atomic clocks.

Italian physicists moved several such 'clocks' on a truck high into the mountains, and several hours later brought them back, to compare their readings with the clock that stayed below all the time. This stationary reference clock was found to lag behind, in complete accordance with Einstein's theory (the difference was measured in nanoseconds, that is, in billionths of a second).

In a similar American experiment an atomic clock was placed in an airplane which was kept in flight at an altitude of nine kilometers for fourteen hours. After landing, the readings of the clock were compared with the reference clock on the Earth's suface. Einstein's theory was again beautifully confirmed.

There is thus no doubt that time is slowed down in the gravitational field. In most of these cases, the changes are almost immeasurably small, but we will see that astronomers and physicists know situations in which the difference between time rates is colossal.

General relativity has completely reshaped our ideas of space and time. Neither is an invariable scene on which the dramatic history of the Universe is acted out. Space is not an infinite rigid skeleton. Moving matter is constantly warping it, changing its geometric properties. The acceptable part of the naive notions of our predecessors about the all-encompassing, immutable time river is gradually dwindling. As we see now, it does not flow everywhere with equal grandeur: it is rapid in gorges but slow over shallows; we will see later how it splits into numerous streams, brooks and rivulets, which move forward at different speeds, depending on 'local' conditions.

7

Holes in space and time

When I began to study general relativity seriously (which was in the late fifties), no one knew well what a black hole would be. Even the term itself did not appear in either strictly scientific or popular science publications. This is a stark contrast with what we see today, when almost everyone has read or at least heard about them. The black hole is a product of gigantic gravitational forces. Black holes are born when the gravitational field, growing in the course of catastrophic contraction of a very large mass of matter, becomes so strong that it ceases to let out anything, even light. An object can only fall into a black hole, pulled by its huge gravitational force, but there is simply no way out.

I first read a description of very strong gravitational fields in the Landau and Lifshitz monograph that I have already mentioned. I studied it while still a student, under Zelmanov's guidance. The book gave a very brief but extremely clear presentation of the properties of the gravitational field of a strongly compressed spherical mass. The solution of Einstein's equations for this case was found by the German astronomer Karl Schwarzschild; consequently, this gravitational field is known as the Schwarzschild field.

I remember that this subsection left me rather indifferent. Nevertheless, I did make some evaluations, using the formulas in the book and the knowledge gained from talking to Abram Zelmanov. I need to repeat that calculations in Einstein's theory are extremely complicated, and quite often the 'forest' of very long formulas hides the physical meaning of the final results. What Zelmanov was teaching me was the basics of this science: clear understanding of the meaning of mathematical derivations. I wish to remark that probably the hardest task in the most complicated of today's theories is to 'distill' the physical message of the results of calculations. I am deeply grateful to my mentor for teaching me the fundamentals of the difficult art of understanding.

I had thus calculated the force with which the central mass

attracts a body located on its surface. The result was rather peculiar. If the radius of the spherical mass was large, the result coincided with the classical Newton's law. But as the same mass contracted to progressively smaller and smaller radius, deviations from Newton's law appeared. Thus, the attractive force grew larger, even though only slightly larger at first. Deviations became quite significant at fantastically high contraction. The most important discovery for me was that for each mass, there was a certain contraction radius at which the force of gravitation became infinite! The theory called this radius the *gravitational radius*. The greater the mass, the greater its gravitational radius, but it is quite small even for astronomical-scale masses: it is a mere one centimeter for the mass of the Earth, and three kilometers for the Sun!

A question immediately formed in my mind: what if the size of the mass is less than its gravitational radius? At first glance, it may seem that the attractive force becomes infinite, but this was absurd and out of the question. Of course, I went to the teacher, who could only say that such bodies are thought to be physically impossible, but that he had never come across an explanation of this assumption. I discovered later that not only Zelmanov, but no one in the world had bothered to look into this matter. The problem was simply left lying off the main road that the science was pursuing. Astronomers did not know of such dense objects in the Universe. Any arguments in this field were regarded as useless and unfounded; moreover, very few people had mastered general relativity at that time. Astronomers assumed that this science was of no use to them: it reigned in the realm of superstrong gravitational fields, while no such fields were observed in the Universe. Nevertheless, I did not forget the impression it made on me, and having become a postgraduate, I chose to tackle the problem seriously.

My first impression was that a body is indeed not allowed to become smaller than its gravitational radius. But I soon realized that I was mistaken; later in the book I will explain the reason for my error.

In 1939 the American physicists Robert Oppenheimer (who later led the team that developed the atomic bomb) and Hartland Snyder gave a detailed mathematical description of what happens to a mass which contracts more and more owing to its own gravitational forces. If a spherical mass contracts to a size equal to or smaller than its gravitational radius, then no external force can stop or reverse further contraction. Indeed, if contraction stopped at the size equal to the gravitational radius, the gravitational force on the surface of the spherical mass would be infinitely strong; since nothing could then resist this force, the mass would collapse further. However, in this rapid contraction, with all matter falling towards the center, gravitational forces are not felt.

We all know that weightlessness sets in in free fall and any body, having lost its support, becomes weightless. The same is true for a collapsing mass: the force of gravity (the weight) is not felt on its surface. Once the body's size has reached the gravitational radius, the collapse cannot be stopped. The mass falls to the center irresistibly. This process is known in physics as *gravitational collapse*; it results in the birth of a black hole. It is inside the sphere of radius equal to the gravitational radius that the gravitation is so strong that even light is not allowed to escape. This region was designated with the term 'black hole' by John Archibald Wheeler in 1967.

The name proved to be eminently suitable and was immediately accepted by all scientists. The boundary of the black hole is known as the *event horizon*. This term is easily understandable because no signal that could bring information about events inside the black hole can cross the horizon and reach the observer. The external observer will never learn anything about processes inside a black hole.

The forces close to a black hole are thus extremely strong, but this is not the end of the story. The reader remembers that the geometrical properties of space undergo changes in strong gravitational fields and that the flow of time is slowed down. The curvature of space becomes very high in the vicinity of the event horizon. We can use the following approach to evaluate the degree of this curving. Replace three-dimensional space by a two-dimensional plane (we remove the third dimension); this will make it easier to picture how it is curved. Look at figure 7.1. The empty space is represented by a plane (a). If we now place a gravitating sphere into this space, the space around the sphere will bend slightly (cave in). Assume now that the sphere starts to contract, so that the gravitational field at its surface grows stronger. This is shown in figure 7.1(b), where the time coordinate, as measured by the observer on the sphere surface, is drawn perpendicularly to the space plane. As gravitation increases, the space curvature grows. Finally, a black hole is born, when the surface of the sphere contracts to a size below the event horizon, and the 'bending' makes the walls vertical. Clearly, the geometry on such a curved surface close to the black hole will be very different from the Euclidean geometry on the plane. We see that from the standpoint of space geometry a black hole indeed resembles a hole in space.

Let us now look at how time flows. The closer we are to the event horizon, the slower time ticks away for an external observer. The tempo dies down completely on the boundary of the black hole. This situation is comparable to the flow of water close to the banks, where the stream almost stops. This colorful comparison was suggested by Professor D. Liebscher when describing the properties of black holes.

However, an observer who departs on a space journey into the black hole on a spaceship will observe a completely different sequence of events. The enormously powerful gravitational field at its boundary accelerates the falling spaceship to a velocity equal to

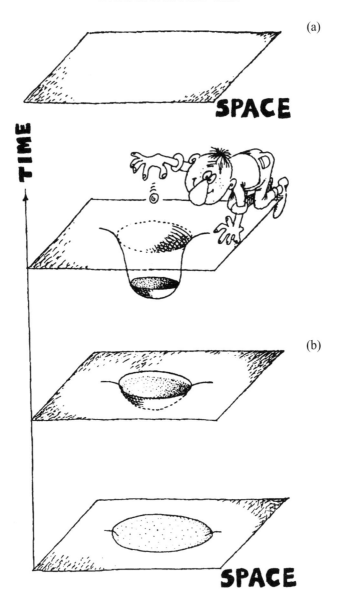

Fig. 7.1.

the velocity of light. Nevertheless, the distant observer is under the impression that the fall of the ship is slowed down and comes to a complete halt at the boundary of the black hole. Indeed, his point of view is that time itself comes to a halt here.

As the velocity approaches the velocity of light, the pace of time on board the ship also gets progressively slower, as it would on any fast-moving object. It is this slowdown that wins over (or rather cancels out) the decrease in the fall of the ship. As a result of this increasing stretching of seconds on the ship, the approach of the ship to the boundary of the black hole extends infinitely but is still measured in a finite number of these stretching seconds (as it seems to the external observer). The fall of the ship that was infinitely long to the distant observer has squeezed into a very short time for the observer on the ship. What was infinite for one has become finite for the other.

This is indeed a fantastic transformation in the perception of the flow of time. What I said with regard to the observer on the ship is equally true for an imaginary observer on the surface of a collapsing sphere in the course of formation of a black hole.

It is probably clear now for the reader where I was wrong in assuming that penetrating a black hole is impossible. I was following this process by the clock of the external observer and concluded that it was infinitely long, when I needed to go by the clock of the co-moving observer. By this latter clock, the fall into the black hole occurs over a finite time, and is very fast.

An observer swallowed by the black hole will never be able to get out, regardless of how powerful the thrust of the engines of the spaceship is. He will be powerless to send out any messages, any signals. Indeed, even the fastest of all messengers in the universe – light – is denied escape. For an external observer, the fall of the ship stretches to infinity by his clock. Therefore, what happens to the inbound observer and his spaceship inside the black hole takes place beyond the time of the external observer (after his infinity in time). In this sense, black holes are holes in the time of the Universe.

I must immediately add a correction: this does not mean that time stops flowing inside the black hole. It still flows, but this is a different time, and it flows in a manner different from the time of the external observer.

When my term of postgraduate studies was almost up, I wrote a paper about this 'different' time, and this is still one of my pet papers. The gist of this discovery of mine was that once we switch from the external space to the space inside a black hole, the time coordinate in the formulas is simply replaced by the spatial radial coordinate, and vice versa. In other words, time transforms into the spatial radial distance, and this distance is time itself!

The reader can reach some degree of understanding of these processes by looking at figure 7.1(b). The space, which was curved when the black hole was being formed, becomes vertical at its boundary (the uppermost surface in figure 7.1(b)); since we plot time along the vertical axis, this means that space (radial direction) is transformed into time. Einstein's theory does predict fantastic things!

When I had written the corresponding equations (the resulting one was only a line long), I immediately, as was the custom, went to show them to Zelmanov. One brief glance at my text was enough; Zelmanov understood everything and a couple of seconds later ordered: 'To be sent to a journal, right away'. This was an over-whelming compliment for a man who usually insisted on months of rethinking and recalculating and honed his own publications for years. This was how I started my research in black hole physics.

At the end of the 1950s, the problem of the inner space of black holes attracted young physicists in other countries: D. Finkelstein and M. Kruskal, then others. However, I was not aware of this at the early stages of my work. I was going forward, and succeeded in finding how a free body would move within the event horizon. I gave this region, in which nothing can be at rest, where everything is forced to fall towards the center, the name *T-region*. This name emphasized the mandatory dependence on time T. Such regions are often referred to in this way even now.

Scientists are known to cherish the terms that they have invented, and feel something akin to jealousy if someone forgets to mention who introduced this precious name. Superficially, this seems strange. Is it not obvious that proving something nontrivial in the theory is a hard job and that anyone who produces a novel result which becomes generally accepted must be much more satisfied with this fact than with the lucky find of a name that took immediate root. In reality, situations are often the opposite of this; I have come across them quite often. I am inclined to think that a scientist is subconsciously influenced by the fact that there is a considerable number of very good theoretical papers while there are only a few good terms that gain universal recognition.

Something very similar happened to me as well. In one of my papers I often cited a well-known Soviet astrophysicist Joseph Shklovsky, who was one of my teachers and senior colleagues; relations between us were quite warm. In this paper, I mentioned the radiation which survived in the Universe from its early age (this topic will be specially discussed later in the book). The term used at that time in publications in the West (and sometimes in the USSR) was rather clumsy – 'microwave background radiation'. Shklovsky suggested calling it the relic radiation. This impressive name appealed to many astrophysicists; as for me, I always used it. Having read my paper, Shklovsky called me and, obviously piqued, demanded why I refrained from mentioning the author of the name. Trying to justify my omission, in complete sincerity, I incoherently babbled that the term had been known for a long time, appeared in dozens of papers, and that citing him about such a tiny thing would be beneath his standing in international science. Neverthe-less, Shklovsky insisted that these small bits were not trivial at all. I think now that he was right. A flashy, emotional name helps concentrate attention on a problem, attracts both young people and mature scientists, and stimulates even those who already work in the field involved. This is a kind of advertising, and anyone in this field knows the importance of a catchphrase. Having invented the phrase 'black holes', John Wheeler helped to popularize the problem of gravitational collapse both among the specialists and among all those not indifferent to scientific puzzles.

Several words now about potential approaches to creating a black hole. Superficially, the problem is not diabolically difficult: take a mass and compress it to the size of its gravitational radius. True, of course, but the snag lies in the gravitational radius being fan-tastically small even for fairly large masses. For example, the mass of a moderate mountain would have to be compressed to the size of an atomic nucleus! It is simply meaningless, therefore, to con-template artificial creation of black holes in today's laboratories or terrestrial laboratories of the foreseeable future.

Even the mass of the Earth would have to be compressed to a sphere of 1 cm in radius; the corresponding size for the mass of the Sun is 3 km.

It was found, however, that nature took care of the job of creating black holes, albeit black holes of very large mass. Such objects may be formed at the final phase of the life of sufficiently massive stars.

I will not touch here on the evolution of stars and their fate at the end of stellar life in any detail. It will be sufficient to say that a star of mass, say, of ten or more solar masses, arriving at the very end of its active evolution, with the nuclear fuel burnt out, will very likely be crushed by the pressure of its own gravitational forces to the size of its gravitational radius and become a black hole. More-over, astronomers can be pretty certain that the first black holes of stellar origin have already been identified. (I will discuss the methods of searching for and identifying black holes somewhat later.) Furthermore, astronomers are sure that central areas of giant stellar systems – galaxies – are the birthplaces of supermassive black holes of masses from one hundred thousand to a billion or more solar masses. Such supermassive black holes may be formed as a result of compression of huge masses of gas that gravitational pressure collects at galactic centers. Another possible mechanism of their formation is the gravitational compression of entire stellar congregations that are found around galactic centers. It cannot be excluded, however, that the Universe also contains black holes of a very different nature.

When astrophysicists developed a serious interest in black holes in the 1960s, physics theorists faced new and very complex problems. Robert Oppenheimer and Hartland Snyder described the birth of the black hole in the compression of a perfectly spherical mass. However, nature never creates an absolutely ideal spherical body. What if a collapsing body is not spherical?

This question attracted me when I had completed my postgraduate term and came to work in professor (Academician) Yakov Zeldovich's group. Our team leader, myself and a friend of mine of the same age, A. Doroshkevich, started to work on the problem. When the work was completed, the result proved to be rather surprising. Calculations demonstrated that the compression of a (nonrotating) nonsymmetrical body produces a black hole which very rapidly becomes perfectly symmetrical. Any deviation from sphericity in the gravitational field must be radiated away in the formation of the black hole and fly away with gravitational waves. The emerging boundary of the black hole – the event horizon – is spherical, and only spherical!

I reported this result at the International Conference on Gravitation in the summer of 1965 in London. This was my first journey abroad, the first serious discussion of the problems involved with Western specialists, the first presentation of results to the international scientific community by the recently formed group of Academician Zeldovich. The debut was successful. It became clear to me that owing to Zeldovich's tremendous physical intuition, his persistence and overwhelming industriousness which fascinated and conquered his students and which was based on his literally childish love of nature's secrets, our small group was leading in a new field of science – 'relativistic astrophysics' (the word 'relativistic' was sometimes a reference to Einstein's theory).

After the report, I was surrounded by a crowd of colleagues who wanted to know details of our calculations. Among these enthusiasts, I immediately spotted a tall, lean, reddish-haired youngster, a typical American, as I imagined. In fact, I had spotted him a couple of days before that: for me, he was the first ever foreign colleague whose brief communication I heard in London. As far as I remember now, he was discussing the gravitational field of a cylinder. His communication was of lively interest to me owing to some sort of similarity, although I could hardly put a finger on it right away, in his and my approaches to problems. After my presentation Kip Thorne (the name of the youth) helped me to make myself understood to people who wanted to discuss my talk (my English at that time was far below the grade of 'far from perfect'). We then went on with our conversation. I quickly established common areas of scientific interest and, which counted at least as much, important similarities in outlook on the world (what a Russian would call 'related souls'). Soon we grew to be real friends. Despite geographic separation (eleven hours difference between the time bands of Moscow in Russia and California in the USA where Thorne lives) and in spite of years without personal contact, the friendship is still very much alive. I now find out from Kip Thorne's book *Black Holes and Time*

106

Warps (W. W. Norton & Co, 1994) that he retained similar reminiscences from our meetings.

Our evaluations of social phenomena were often very similar, the same women looked attractive to both of us, and it was almost self-evident that both would start working simultaneously on the same problems. During one of my visits to Thorne's laboratory (he is Professor at California Institute of Technology), I was giving a talk at his seminar and offered criticism of the proof of a theorem in one recent publication by another colleague. I had just opened my mouth to formulate my arguments when Kip, who had also read that paper, stopped me: 'Igor, I know what you are going to say'. We compared our arguments: they were identical to a tee, and this was uncanny. Not yet back to my normal composure, I weakly asked him, could he explain such coincidence in thoughts? Smiling Thorne replied: 'We have the same background, almost a quarter of a century of acquaintance and detailed knowledge of one another's work'. (Kip edited the translations of the astrophysics books written by Zeldovich and myself, while I edited the translations of his books published in Russia).

The tribute to Kip in the paragraphs above was written many years ago for the Russian edition. Now, in the English edition, I want to outline, very briefly, the events in our friendship in the years after the Russian edition had been printed. I relish this chance, and the reminiscences may cause unfamiliar agitation and pride for people in general and for the fraternity of astrophysicists in particular.

Here is my story. When staying as a visiting researcher in Copenhagen in 1990, I had a microinfarction. Cardiologists recommended detailed heart inspection. Back in Moscow, I went to specialists. They confirmed the seriousness of the predicament but, with a sigh, confessed that they could not recommend – in the USSR – even the required special tests, let alone a full-scale surgical treatment for heart vessels: 'Too dangerous. Our surgeons are wonderful but as for the rest... Statistics show that life expectancy for post-operational patients is no longer than for those who were not operated on at all' was their advice. I knew already that well known theoretician E. M. Lifshitz had recently died after a similar operation.

What was I to do? I definitely had no hard currency for heart surgery in the West. My wife wrote to Kip. Kip immediately arranged for everything with the best experts in Los Angeles, so that only a month later, after another conference in the USA, I was in Kip's house in Pasadena (California). Testing demonstrated that the operation was to be considerably more serious than was predicted back home and that it must be done urgently. Kip dropped everything he was doing, and together with Carolee, his wife, never left my side. Even though the cardiologist, the surgeon and the leading anesthetist of the Huntington Memorial Hospital in Pasadena decided to forego the treatment fees, Kip's personal means were insufficient for the operation to go on. He wrote to physicists and astrophysicists who knew me. As a result, dozens and dozens of colleagues, physicists and astrophysicists, sent money to cover the cost of the operation. In the meantime, the operation was successfully carried out. When consciousness began to return to me after the operation that lasted for hours, the first thing I understood was Kip's voice repeating (in Russian!) 'Igor, everything is fine, just fine!' Even though my daughter, with whom I was always very close and who did very much for my recovery, was sitting by my bed, I perceived as something self-evident that it was the voice of Kip Thorne which brought me back to life.

It is said that the great physicist Lev Landau, the victim of a terrible automobile accident, who was returned to life (unfortunately, not to physics) by an incomparable effort of doctors, physicists and simply admirers of his talent, mused after his resurrection: 'It is rather a pity that I used to be of a much lower opinion of people than I am now'. I cannot say that my view of humanity was particularly low before this chain of events in my life. I thought, nevertheless, that I dispassionately judged the tendency of people to concentrate almost exclusively on their own worries and to be guided by a logic of the type 'life is hard and complicated, you cannot take care of everyone'. I was wrong. I am happy now to have discovered that people are much above the 'averaged' criteria that I regarded as accurate.

The wonderful friendship of Kip and his wife (who make a movingly harmonious couple), the unselfishness, exceptional skill and attentiveness of American doctors, and finally, the brotherhood of the physics and astrophysics community, granted me a second life. My health is excellent, I enjoy my favorite sport (water-skiing), and – above all – I do physics. My thanks to the world for that.

Let us return to the sixties. Three years after the London conference, Kip visited the USSR to attend the International Conference on Gravitation in Tbilisi (now Republic of Georgia). He told me that despite the lively interest caused by our work on the collapse of a nonspherical body, not all Western theorists accepted the most important conclusion that the compression of a nonspherical body also produces a black hole. Among these skeptics was a very well known scientist Werner Israel. Thorn indicated that the doubts stemmed from our assumption that small deviations from sphericity cannot grow to infinite amplitude when the body collapses to the size of the gravitational radius. The physical intuition implanted by Yakov Zeldovich prompted one to believe that this assumption is quite obvious. However, mathematicians demanded proof, and I started working on it.

A year later, when Kip Thorne came to the USSR again and was leaving after six weeks' work with our team, I was able to send with him the completed paper with the demanded proof. As far as I know, it was universally accepted.

In this paper I showed that if the surface of a spherical body is slightly 'rippled', and the body collapses to the gravitational radius, the 'ripples' slightly grow in the process but stay small and do not tend to 'blow up' to very large size. What our first publication lacked was precisely the proof that the ripples remain small. I thought up a mathematical proof which was very simple; it seemed almost trivial to me. To my surprise, our colleagues in the West found it to be seminal. Probably, the proof turned out to be relatively simple because I was familiar with the research on the mathematical methods of constructing the so-called 'general solutions' in Einstein's theory. These constructions were developed by the Soviet physicists (later Academicians) Evgeny Lifshitz and Isaac Khalatnikov. I also knew well the work of the Soviet mathematician Aleksei Petrov, so that I only needed to modify and extend these ideas and apply them to the problem I needed to solve.†

This example shows once again how important it is to follow the progress in 'contiguous' fields of science.

My story is now in the late sixties, when mutual visits of Soviet physicists to the USA and of American physicists to the USSR were much less frequent than nowadays. Each trip was an important event, discussed in detail at scientific seminars. The reader has to recall that at that time our country had no fax or e-mail facilities, no electronic networks for spreading information, and that all our telephone conversations with our Western colleagues were strictly controlled. We were coming back from our voyages with fresh information on what our colleagues were working on. It was at least as important that we were learning how things were done, were familiarizing ourselves with the new style of research, with methods that often differed greatly from ours. Nothing can harm science research more than isolationism, absence of contact with the global scientific community, the impos-

---

† The complete theory of black holes was constructed by the effort of a large number of theorists. However, it is not my task here to describe the entire story and list all the names.

sibility of communicating with colleagues frequently and thoroughly. A non-scientist can hardly realize the degree to which research is stimulated by constant discussions, exchange of opinions, even simple contact with colleagues coming from other scientific schools, representing different techniques and approaches (I assume that these are colleagues at the forefront of scientific research).

In February 1967 I returned to Moscow from my first trip to the USA to the so-called Texas Symposium on Relativistic Astrophysics (called 'Texas Symposium' because of the place where the first of these symposia was convened). This time it took place in New York; the second symposium in this field, it reflected the greatly changed situation in theoretical and observational astrophysics.

A large number of astrophysicists were aware that nature is hiding bodies that differ dramatically from anything that astronomers had studied before. These objects must be extremely unlike ordinary stars, or planets, or rarefied gas. These hypothetical bodies generate enormous gravitational fields which are described by general relativity. This is why they were called 'relativistic objects', and why the symposium was given this name. The talks were devoted to as yet undiscovered neutron stars and black holes.

The delegation of the USSR Academy of Sciences consisted of only three people: Vitaly Ginzburg, Joseph Shklovsky and myself, among something like several hundred participants. Despite our desperate efforts to gather maximum information and talk to as many colleagues as possible, we were physically unable to cover all that was of interest. Even though this was almost thirty years ago and many things have changed for the better, delegations of astronomers and physicists from Russia to international gatherings continued, for the subsequent decades, to be smaller than delegations from the USA, and even of much smaller (and less developed) countries, by a factor of several tens (and sometimes several hundreds!). This so-called 'money-saving' policy was very harmful to science in general (our physicists and astronomers are leading in many fields). In recent years (in mid-1990s) many scientific visits from the former Soviet Union are made possible only by financial sponsorship from the West.

After the symposium, we were invited to different research centers. I went to Princeton, to the Institute of Advanced Studies, where Albert Einstein spent the last decades of his life. Together with Kip Thorne, we were guests of John Archibald Wheeler and lived in his house (Thorne had been Wheeler's student). This close contact with these physicists of very different generations proved that the problem of searching for relativistic objects is treated as very serious in the West.

I wish to emphasize that organizing a search for relativistic objects in the Universe was pioneered in 1965 by Yakov Zeldovich and Oktai Guseinov, then a very young astrophysicist from Azerbaijan who joined our team. The difficulties in searching for such objects are as follows.

Theory knew at the time of two types of relativistic bodies: neutron stars and black holes. The size of a neutron star is only about ten kilometers, so that they emit very little light even if their surface is very hot. As for black holes, it was assumed that they emit absolutely nothing. Hence, we cannot hope to detect either of these objects at the great distances separating us from these heavenly bodies.

How should we go about looking for them?

The Soviet physicists Vladimir Braginsky and Aleksander Polnarev once jokingly remarked that a discussion of the problem sometimes sounds very much like the conversation between King and Alice in Lewis Carroll's *Through the Looking Glass*:

'Just look along the road, and tell me if you can see either of
them.'
'I see nobody on the road,' said Alice.
'I only wish I had such eyes,' the King remarked ... 'To be
able to see Nobody! And at that distance too!...'

In fact, Zeldovich and Guseinov noticed that invisible relativistic
objects will generate enormous gravitational fields. They will be
detected precisely through these fields. In their opinion, relativistic
bodies must be sought among binary stellar systems in which the
gravitational pull of the invisible component influences the motion
of the neighboring star. The presence of the invisible companion
can be deduced from the peculiarities in the motion of the visible
star.

After his acquaintance with the work of Zeldovich's team, Kip
Thorne got excited with the idea of finding relativistic bodies in the
Universe. This is one example when research in the USSR stimulated
American physicists. Together with Virginia Trimble, Thorne published a catalogue of stars sufficiently suspicious for their neighborhood to hide invisible companions with strong gravitational
fields. Alas, detailed studies of the stars in this list and stars indicated by other astronomers did not lead to discovery of relativistic
heavenly bodies.

Neutron stars were discovered accidentally in 1967 by English
astronomers, who noticed their characteristic radio emission.

The discovery of black holes was lagging far behind. In 1966 Zeldovich and myself, and in 1967 Shklovsky, pointed out that black
holes (and neutron stars) can act as extremely powerful sources of
x-ray radiation. This situation will occur if an ordinary (normal) star
exists quite close to a black hole. According to the picture generally
accepted now, the gravitational pull of the black hole will force the
gas from the atmosphere of the normal companion star to flow in
a spiral towards the black hole, forming a compact gas disk. As a

result of friction between gas layers, the gas in the disk heats up to a temperature of tens of millions of degrees and, before it sinks into the black hole, emits x-rays. This x-ray emission makes the black hole visible. X-ray sources in binary stellar systems were first discovered in 1972. Some of them proved to be neutron stars. The other sources, in the opinion of the majority of experts, are black holes. To those readers who are interested in the history of the idea of black holes and of their astrophysical discovery, I recommend reading an excellent article by Werner Israel 'Dark stars: the evolution of an idea', in *300 Years of Gravitation*, ed. S.W. Hawking and W. Israel (Cambridge: Cambridge University Press, 1987, p. 249).

Not long before these events, I met the English scientist Stephen Hawking who later became one of the most outstanding physics theorists of our century and, without any doubt, the greatest expert on black holes. Hawking's name is now familiar to anyone even slightly interested in physics by virtue of his best-selling book *A Brief History of Time*. A number of books and numerous articles in papers and magazines have been written about him. I have no doubt that his name is well known to the reader of this book (indeed, it was an interest in physics and astronomy that attracted you to the book title, right?). Nevertheless, I will briefly outline my impression prompted by his scientific discoveries and by our encounters. We met in 1972 when I attended the General Assembly of the International Astronomical Union in Brighton on the south coast of England. A young British astronomer Malcolm Longair, future Astronomer Royal for Scotland, who spent considerable time training in Zeldovich's team, invited several Soviet delegates to visit the Astronomy Institute and the famous radioastronomical observatory in Cambridge. It was in this observatory that a young research student S. J. Bell and her supervisor A. Hewish had discovered, three years previously, the neutron star, by detecting its pulsed radio emission.

I looked at the unusual radiotelescope, with which the discovery was made, with genuine and undisguised curiosity. This was a large field (four acres in size) with poles dug into the ground and horizontally stretched wire between the poles, forming the antennas. The telescope was designed and constructed by Tony Hewish. The bulk of the construction was carried out by assistants at the Observatory, and students were involved in putting it together. This curious 'machine' made possible the discovery of heavenly bodies whose gravitational field was so intense that breaking this bond demands velocities almost equal to the velocity of light.

Neutron stars proved to be a testing ground for studying numerous puzzling phenomena. For example, their magnetic field is so strong that each cubic centimeter of space at their surface contains an amount of energy equivalent to a hundred grams of mass! This density is a hundred times greater than that of water, and is greater than the density of any mineral or chemical element under natural conditions. Note that this is merely the magnetic field which we habitually treat as something quite ephemeral.

As a result of the strong gravitational field, time on the surface of the neutron star progresses one and a half times more slowly than in our world, but at the center of the star the slowdown factor reaches two and a half.

Even on the way to Cambridge, I arranged with Longair to visit Stephen Hawking at his home. It could only be at home because Hawking was already suffering from a severe illness, the amiotrophic lateral sclerosis (ALS), known as 'Lou Gherig's disease'. This disorder attacks the central nervous system, leads to gradual atrophy of muscles and several years later kills the patient. Hawking fell prey to this disorder in his early twenties when he was completing his doctoral thesis. No wonder that he felt crushed by the news of the progressing illness; he saw no point in completing the Ph.D., threw his science out the window and was drinking heavily. Fortunately, the development of the disorder slowed down, and then fate made him a gift: he met a charming young lady, Jane, to whom he was soon engaged. This was the turning point of his life. Later he would remember: 'If we were to get married, I had to get a job. And to get a job, I had to finish my Ph.D. I started working hard for the first time in my life. To my surprise, I found I liked it.' (*Time* February 8, 1988).

By the time I arrived in Cambridge, Stephen Hawking was already widely known for his analysis of the initial stages of the expansion of the Universe (we will take up this topic later). Hawking proved that the Universe began expanding from a state of extremely high density and extremely high gravitational field; physicists say that it began with a singularity. I was looking forward to meeting him with great impatience.

Hawking has always immensely impressed people who have never met him before. Several minutes are enough to forget that you are near a very ill person who is virtually unable to move. His eyes are fascinating in their luminosity. One immediately senses the infinite profoundness of his intelligence and never ceases to delight in the magnificence of these eyes.

It was not easy for me to understand his words since he spoke with great difficulty but Longair, who had talked with Hawking for some years, helped me to interpret his words. I described to Hawking what Zeldovich and I were trying to achieve in Moscow in that period. For some reason, I did not expect Hawking to be interested in the mathematical details of the work and thus remarked that I would omit them. But he smiled and countered that these details are the most important thing. At that time we both devoted much time to cosmology. From this conversation I felt that his focus was definitely moving towards black holes. As for myself, I have always believed that the key to many deeply buried mysteries of nature is tied to black holes.

Fortunately, Hawking's disorder has stabilized with the passage of years. In spite of the initial grave pronouncements of doctors (and obviously owing to their care and Hawking's mental strength) he is alive and working. I feel that his intellect is growing more profound, even though he has lost control by now of almost all his muscles. He can move only on a special wheelchair that he controls electronically, by the fingers of the left hand. His power of speech is gone, so he talks to people only through a computer. Nevertheless, he is still full of humor, is active and jolly, participates in excursions, goes to theaters and restaurants, invites people to his house, is always surrounded with people.

But the most important thing for us is that he works as no one else. The science world bows to his profound ideas which he keeps generating one after another. They are always extraordinary.

Hawking visited the former USSR a considerable number of times. His last visit was in 1988. He came to the International Conference in Leningrad (now St Petersburg) which commemorated the anniversary of the birth of Aleksander Friedmann, the creator of the theory of the expanding Universe. Hawking gave a talk to the conference, took part in a number of excursions and traveled all over the city. During the conference, I recorded an interview with him for Soviet television.

Hawking has three children: two sons and a daughter. Incidentally, Hawking was born on January 8, 1942, exactly three hundred years after the death of Galileo (he often mentions this himself). This is how he describes the gross outlines of his life:

> Apart from being unlucky enough to get ALS, or motor neurone disease, I have been fortunate in almost every other respect. The help and support I received from my wife, Jane, and my children, Robert, Lucy and Timmy, have

made it possible for me to lead a fairly normal life and to
have a successful career. I was again fortunate in that I
chose theoretical physics, because that is all in the mind. So
my disability has not been a serious handicap. My scientific
colleagues have without exception been most helpful.

Our next meeting after the first encounter was in 1972. Several
invited speakers, Hawking and myself among them, were giving
lectures on black hole physics at the International School in Les
Houches in the French Alps. Hawking came with his charming wife
and two kids who at the time were quite small. Jane did not forget my description of my son's passion for toy cars and brought
specially for him one such tiny car (a wonderful rarity in the USSR
then). At that time, Hawking was still lecturing in his proper voice,
even though speaking was already a great effort. He dictated the
main ideas of the talk in advance to his assistants and they would
show them to the audience during the lecture, with Hawking only
clarifying these statements.

We would often come together of an evening in the cozy halls of
the school, to talk of science and life. Hawking reminisced about
how he loved figure skating in his youth. It was quite painful to
see how cruel fate was to this vigorous, smiling, witty man. Even in
this state, Hawking's vigor was ahead of even the fiery young Italian
professor R. Ruffini, who also lectured at Les Houches. Our lectures
were later published as a book which became the first comprehensive exposition of black hole physics and was to be the starting
point for numerous further studies.

An important feature of the new approach to the black hole problem was that they were no longer treated as something belonging
to a physicist's graveyard: 'gravitational graves', into which matter
can only sink and thus disappear for an external observer. As a matter of fact, the powerful gravitational field of a black hole interacts
with the surrounding medium and generates violent physical pro-

cesses. Ruffini used to say that black holes are 'very much alive'. For instance, the heated gas flow in the disk, spiraling around a black hole that is a component of a stellar binary system, must produce intense x-ray emission.

I have mentioned already that the first x-ray sources in stellar binaries were discovered in 1972. The parameters of one of them, Cygnus X-1 (in this designation, X stands for x-ray, Cygnus is the name of the constellation in which the source was discovered, and 1 is the number of the source) indicated that it contained a black hole of a mass of about ten solar masses.

Of course, we spent considerable time at Les Houches discussing the new discoveries. Indeed, what seemed imminent was exposing the holes in space and time. Our group split into optimists and skeptics. Optimists firmly declared that the discovery was a fact. Skeptics insisted on caution and demanded rechecking of facts.

As for me, I accepted this discovery with all my heart. In hindsight, I rather think that my intuition did not lead me astray, even though I am getting more cautious as the years go by. Thorne, who also was one of the invited speakers at Les Houches, held a similar view but believed that the observations needed additional verification. Two years later, new information was gathered about the source in the Cygnus constellation. Kip Thorne wrote that the new data made both him and many other astronomers accept that a black hole does lie at the center of Cygnus X-1.

Stephen Hawking was much more guarded. Remembering those years, he wrote in 1988:

> A black hole seems to be the only really natural explanation of the observations. Despite this, I have a bet with Kip Thorne of the California Institute of Technology that in fact Cygnus X-1 does not contain a black hole! This is a form of insurance policy for me. I have done a lot of work on black holes, and it would all be wasted if it turned out that black

119

holes do not exist. But in that case, I would have the consolation of winning my bet, which would bring me four years of the magazine *Private Eye*. If black holes do exist, Kip will get one year of *Penthouse*. When we made the bet, in 1975, we were 80% certain that Cygnus was a black hole. By now, I would say that we are about 95% certain, but the bet has yet to be settled.

The wager was indeed done according to the strictest rules and even appeared in an official publication. For the reader to better grasp the humor of the bet, I should probably indicate that the magazines in the bet are light-years away from the scientific press. Seriously, though, I believe that Hawking's estimate of the reliability of the black hole discovery news was fairly plausible. I would say now that my estimate of the degree of certainty is closer to 100%. Of course, astronomers are prudently cautious because what is involved is not the discovery of just another heavenly body but an announcement of the existence of holes in space and time.

This is not yet the end of the story. The facsimile with the Thorne–Hawking bet, published in 1987 in Werner Israel's paper in *300 Years of Gravitation*, mentioned earlier, was accompanied with the following caption: 'A bet between Stephen Hawking and Kip Thorne made at CalTech in December 1974, on which neither side has yet collected'. In late autumn 1991 I phoned Thorne and said this: 'Kip, don't you think it's time for you to collect on the bet? Nearly twenty years have passed since the discovery of Cygnus X-1. All facts point to it being a black hole and nothing tells us otherwise. Other black hole candidates have been found. Enough caution! It already resembles the philosophical attitude of a solipsist who doubts the existence of the world around him because it is not possible to prove with pure logic that the external world is not merely a picture created by individual conscious mind.'

Kip laughed and said that I'd laugh too because he had already collected! Hawking had mailed him the subscription, and Kip's secretary was mighty shocked when the magazine started to come to the office.

I was very much surprised: 'Do you mean that Stephen declares that black holes can be treated as having been discovered and confirms this attitude 'officially' in this unorthodox manner?'

Kip guardedly remarked to this that he could not be sure but that it was very likely since he had a concessionary note on the bet, written on Hawking's behalf. Kip was always meticulously attentive to anything concerning the words of other people.

'Fine,' said I, 'but how about you? Do you consider that black holes have indeed been discovered?'

Kip answered in the same guarded manner of a true theoretician that no new data of principal importance had appeared on Cygnus X-1 in the last, say, three years; hence, his estimate of a black hole presence in this object was, as before, around 95%. However, if we talk of the hypothetical presence of a black hole among Cygnus X-1 and the new black hole candidates, he was ready to considerably raise the figure.

'Okay', said I, 'I am preparing now the English translation of my popular-science book *The River of Time*. Will you send me, in writing, your today's estimate, to be placed in my book and thus become engraved on history's plaques?'

Here is the text that arrived from Kip:

> I would estimate probability of a black hole in Cygnus X-1 to be 95%; probability that some of the many other black hole candidates do contain black holes, 99%.
> Best wishes
> Kip
> PS: but don't ask me WHICH other candidates are really black holes; I don't know!

Isn't this a proper place to recall the advice given by young Isaac

Newton to his friend Aston (I mentioned this letter on pp. 9–10): to ask questions and express doubts during travel but to avoid definitive statements and shy away from debates.

We can now return to the physics of these fascinating objects. We will be mostly concerned with how time flows inside them.

I have mentioned already that for an external observer, time flow at the boundary of the black hole slows down just as water flows sleepily by river banks.

It may seem that we should be indifferent to what happens inside the black hole. Indeed, we cannot look into it, nor extract any information out of it. Thus it appears that the inside of the black hole is separated from our Universe by an impenetrable barrier. In fact, this conclusion is only half true. The black hole boundary is semi-transparent: falling into it is allowed while getting out of it is forbidden. What is to happen to an observer and his spaceship after they fall into a black hole? We know that they cannot escape: the irresistible gravitational force pulls them into the heart of the black hole. What is their fate?

Theorists assumed, even not so far back, that having passed through the 'gorge' of the black hole, the observer might reappear from another 'gorge' in our space, far from the black hole into which he fell (figure 7.2(a)). Or he may even 'pop out' in the space of a different Universe (figure 7.2(b)).

If this were possible, the Universe would contain, in addition to black holes, also 'white holes': those other gorges that 'regurgitate' the observer. An object can only be ejected from a white hole, but is forbidden to fall into it. Black and white holes resemble one-way streets for city traffic, with the gorges often referred to as tunnels. But these are streets through time!

I discovered white holes in 1963 by pure mathematics, when trying to find the source of gigantic amounts of energy released in quasars (quasi-stellar radio sources; they are the extremely power-

(a)

(b)

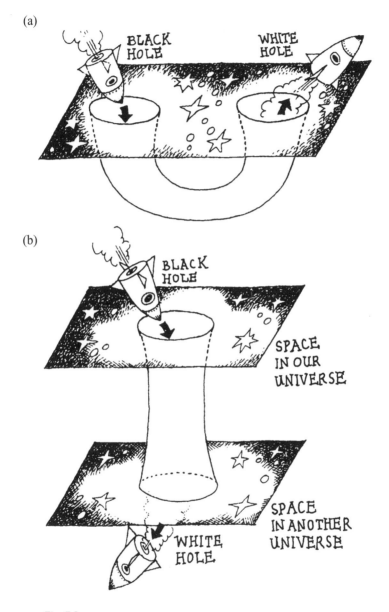

Fig. 7.2.

ful emitting nuclei of some galaxies). A year after I had published a paper on white holes, they were rediscovered by the well-known Israeli physicist Y. Ne'eman who made important contributions to the theory of elementary particles. Soon, however, Ne'eman turned to very different matters and became an important political figure of his country. As a result, I never had a chance to discuss the problem of white holes with him.

The problem of 'tunnels' connecting holes was formulated (long before I published the work on white holes) and discussed by John Wheeler and his students. I will return to this aspect in the chapter 'Against the flow'.

My work of that period on the theory of 'gorges' appealed to Professor Andrei Dmitrievich Sakharov†, who was getting more and more interested in problems of gravitation and cosmology. Now the name of this illustrious personality of the 20th century is familiar to everyone, but at that time he was known only to a very narrow circle of physicists engaged in nuclear weapons research. At that period Andrei Dmitrievich Sakharov was developing cosmological theories which partly overlapped the fields in which I worked. We were discussing these subjects among ourselves and with Yakov Zeldovich. As a result, we published, together with Andrei Dmitrievich Sakharov, our papers as a preprint of the Institute of Applied Mathematics *Relativistic collapse and the topological structure of the Universe* (1970). I am very proud of this work. In his paper Andrei Dmitrievich considered compression of matter resulting in formation of a black hole and its subsequent expansion in a 'different'

---

† A. D. Sakharov (1921–1989) – outstanding Soviet physicist, 'father' of the Soviet hydrogen bomb; in the 1960s, 70s and 80s, having left military research, actively championed end of confrontation with the West and fought against violation of human rights in the USSR; was persecuted by the KGB and exiled to the city of Gorki from 1980 to 1986. Received Nobel prize for Peace in 1975. *Translator*

Universe, as he put it, 'after I. Novikov'. Soon after this publication, Sakharov acted as an official referee of my D.Sc. dissertation.

I will later return to Sakharov's ideas on the nature of time. Here I wish to add a few strokes to his portrait.

I first met him in 1963. His appearance and manners least resembled a great scientist or even an expert physicist. I have seen quite a few prominent physicists since then, and they are all very different people. For instance, my mentor Abram Zelmanov was a somewhat prim, slow and graceful person, both in gestures and actions – the classic bookish picture of the old-school scientist. The Italian Remo Ruffini, an explosive southerner, delivered his lectures in 1972 at Les Houches 'in full swing', presenting his ideas on 'living black holes' with fire in his eyes, his shirt drenched in sweat, covered with chalk and jumping from one corner of the blackboard to the other. Kip Thorne impressed with his wonderfully relaxed manner and the exceptionally precise wording of his talks. I remember what seemed to me excessive histrionics in the talk of Murray Gell-Mann when I first heard him in Philadelphia in 1967. Richard Feynman's presentation was always irresistibly artistic and charming; during our visit to Disneyland he drummed a tune on a bollard and looked perfectly identical to his famous photographs with a drum: very unusual for a physicist, let alone a Nobel prize winner.

I think, nevertheless, that all this variety does not prevent physicists from recognizing a colleague 'at first sight'. A Russian proverb says: 'A fisherman tells another from afar'.

This was definitely not true for Andrei Sakharov. On the day of our first meeting, he knocked rather timidly on the door of the Department of Astrophysics (the department existed in a single room), recently created by Yakov Zeldovich in the Division of Applied Mathematics of the Mathematical Institute in Moscow. The 'Division' was headed by Mstislav Keldysh, then President of the Academy of Sciences, and was in itself a huge research institute. It dealt almost exclusively with strictly classified work ('closed topics', in Russian parlance) for the military establishment. Our astrophysics department, having no direct link with the rest of the institute, was created, on the one hand, because Zeldovich, a very 'classified scientist', longed to start work in this field. On the other hand, Keldysh was of the highest opinion of Zeldovich's work and was happy to invite him to the institute.

I will continue the story of Sakharov's visit. At the end of that working day I saw coming into the room a tall, thin stranger in a fairly worn-out and not very clean winter coat. I was glued to my calculations, my thoughts were far from the mundane surroundings and I was rather irritated by the distracting visit. The unknown visitor looked strange, and the strangest thing about his face was his eyes. Before my recent move to Zeldovich's group, I had worked for a short time at the Astronomical Institute in Moscow, a completely 'open' (not classified) establishment, the ultimate opposite of the one we were in at that moment. The Astronomical Institute was a place that attracted amateur astronomers; they used to bring for review their homespun hypotheses that offered 'keys' to the origin of the Solar System and sometimes of the Universe as a whole. At best, the hypotheses were based on high-school physics. In the majority of cases, these people did not have the slightest idea of what science is about but were invariably sincerely engrossed in their ideas, often to the degree of obsession.

Something in the appearance of Andrei Dmitrievich Sakharov, and especially his eyes, reminded me of these unwelcome visitors. I do not want the reader to think that there was something 'abnormal' in his appearance. Not at all! But the way he looked at you made you suspect that he saw much more than just the real world around us. I understood later that the only trait that had reminded me of amateur astronomers was this extremely 'unconventional' expression in his eyes and of the entire face, plus a certain casualness of his dress. I would even dare to say that the way he looked at you was as extraordinary as that of Leonardo's Mona Lisa, even though there was no hint of the enigmatic smile.

I have said already that I was annoyed by the break in concentration, by the prospect of tedious conversation with another lover of lightweight hypotheses; I had even forgotten for a moment that no amateur would be allowed into our 'sealed' institute and so responded to a mild question if this was Zeldovich's department by demanding that the visitor first return to the ground floor, check his overcoat and only then return. 'Yes, of course' replied the visitor in his now universally known soft voice, and left. When the door closed after him, my peer and colleague Andrei Doroshkevich, who was acquainted with Sakharov at his previous work in an utterly classified place, expressed his disgust: 'Are you crazy? Is this the way to treat Sakharov?'

I was really shocked. Because of the secret nature of his research, I knew very little about his scientific achievement, and most of his heroic social human-rights activity, which made of him a symbol of conscience and invincibility, was still in the future. But even the little that I knew was sufficient to be in awe of his results, and my blunder made me blush.

After this first occasion, Sakharov often paid visits to our laboratory. His motive was to talk physics with Yakov Zeldovich, but Doroshkevich and myself, still very young collaborators of Zeldovich, were also present (but mostly kept our mouths shut). Most of what Sakharov discussed was hardly known or understandable to us. To me (and very likely to others) the subject seemed extremely abstract and far removed from any conceivable reality. Today the world knows how uncompromising Sakharov was in the crucial matters of social life; in fact, a similar independence of any outside pressure could be felt in the evolution of his scientific ideas. It seemed to me that in the inner development of his ideas

127

he never accepted any 'compromise' that might 'smoothen' their unconventionality. His approach to solving scientific problems was more than just unconventional.

For example, Sakharov suggested theoretically calculating the coefficient of universal gravitation – one of the few 'fundamental' constants of nature – from the assumption that it describes the 'elasticity of the vacuum'. Odd, isn't it? However, from the standpoint of today's physics (decades after Sakharov's work) this idea does not appear to be all that outlandish.

I do not know whether this unusual approach to scientific problems is fruitful or not, but there can be no doubt that no one else could work in this way. Perhaps Sakharov resembled Einstein in this respect but this is something of which I do not risk to offer an opinion.

When Sakharov progressed from elementary particle physics and field theory, in which I was not very knowledgeable, to the physics of black holes, where I felt quite strong, the possibility of understanding increased considerably. However, even then the logic of Sakharov's inferences often baffled me. In fact, not only me. The following story comes to mind.

Andrei Sakharov was giving a talk at our seminar about the possible properties of the 'gorges' in space (we have discussed them above in this book). By that time there were about ten young colleagues in our group. We listened in silence. Yakov Zeldovich always treated this topic, to put it mildly, with disapproval. Furthermore, he was annoyed by Sakharov's quaint approach to treating the problem. To some extent, 'strangeness' (of Sakharov's approach) was multiplied by 'strangeness' (of the topic itself). Zeldovich felt that this produced a 'perversity' in physics. Some time after Sakharov began his talk, Zeldovich started to show signs of impatience. Nevertheless, he did not interrupt the speaker even though questions and debates during presentations were a matter of course at our seminars. Quite often, heated debates flared up in the middle of a talk, and lasted longer than the talk itself. The unusual thing this time was rather the silence of the audience, and above all Zeldovich's silence. At last he stood up, went behind his chair and leaned on it. Sakharov looked at him inquiringly but Zeldovich merely waved him on: don't pay attention to me, go ahead with the talk. Sakharov completed the presentation; 'Is that all?' asked Zeldovich. 'Yes, this is all I planned to describe' replied Sakharov softly, and blinked. Zeldovich immediately asked for a newspaper, carefully spread it on the floor and, under our unbelieving stares, knelt on it in front of Sakharov. Placing his palms in a position of worship, he exclaimed: 'Andrei Dmitrievich, please stop doing this nonsense!'

The relationship between Sakharov and Zeldovich was not always that warm. Zeldovich sometimes disapproved of Sakharov's actions in his human rights campaigns. Speaking at Zeldovich's funeral service, Andrei Dmitrievich said: 'We had our periods of friendship and periods of alienation. We had what we had...'. I would like to emphasize, however, that even though Zeldovich may have disagreed with some specific scientific ideas of Sakharov, he always recognized Sakharov's greatness as a physicist. Andrei Dmitrievich was equally in awe of Zeldovich's genius.

Before returning to the story of white holes and gorges, I want to add several more words about Sakharov. Both in science and in his social role, he appeared to me, if anything, as a kind of saint (at least this is his image in my memory now) who hovered above the mundane sides of many a daily problem (mundane to him, not to us, common mortals); not in the sense of being indifferent to them but rather of looking over and beyond them and seeing where we ought to go. I never saw another person, in science or life, who would possess more inner calmness, firmness and certainty than Sakharov. In all likelihood, the extraordinary expression in his eyes with which I started the story of first meeting him, could be best characterized by saying that you looked into the eyes of a saint.

These impressions are in resonance with the words of my professor and older colleague Joseph Shklovsky said about an American colleague of Andrei Sakharov who, like Sakharov, took part in developing nuclear weapons, only on the American side:

> I mean Phil Morrison, currently one of the leading American astrophysicists. Seriously ill, almost an invalid, Morrison realized even in the 1940s that a scientist's honesty and honor are incompatible with serving the gods of war. Morrison resigned from the Los Alamos lab with a scandal and banged a door hard. The consequences were very serious for him but he was never crushed. Sitting with him at a table in a Mexican restaurant in the older part of Albuquerque, some hundred miles from Los Alamos, I looked into his blue, utterly childish, clear eyes – eyes of a man with crystal-clear conscience. It warmed my soul.
>
> J. S. Shklovsky, *Echelon*, 1991

Can we single out Sakharov's statements that would be not just incomprehensible to us but plainly erroneous? To anyone with a feeling of what science is, this question is strange. Any scientist, however successful, inevitably makes mistakes. However, most people who are far removed from the realities of searching for the scientific truth firmly believe (often unconsciously) that being a genius means understanding everything and uttering only ultimate truths. This image is definitely wrong. A scientist, even a genius, is not a prophet. He has to make mistakes in search of truth, otherwise the truth will not be found. A search for truth is a hard toil of non-interrupted deliberation, calculations and resumed reflection. This is torture and joy at the same time. Errors are thus simply unavoidable. You do not find in science a revelation in, say, its pure form. Goethe said: 'He who seeks cannot but roam'.

Yes, Sakharov committed errors too, but as everything connected with him, these errors were unusual.

Once, in a discussion about the fate of ordinary matter and black holes in the future Universe, Sakharov said that sooner or later, all matter will some day fall into black holes spread through the Universe. I virtually let my jaw drop to the floor. Zeldovich was also baffled, he said: 'Andrei Dmitrievich, the Universe is expanding, all bodies recede from one another, the probability of collisions of both heavenly bodies and of atoms of intergalactic gas with a black hole constantly decreases and only very few have a chance of being absorbed by a black hole.'

Sakharov parried with a joke: 'I cannot follow the fate of each individual particle, as a general cannot follow the fate of each of his soldiers. I trace the essential strategic line.' A joke was his typical response in a situation like this. The Moscow theorist Lev Okun recalled one such joke:

> I talked with Andrei Dmitrievich Sakharov about a paper
> which M. B. Voloshin, I. Yu. Kobzarev and I had submitted to
> a journal. The basic idea of the paper was that a vacuum can
> be unstable. For example, in our work a vacuum is capable
> of converting spontaneously into another, stabler state
> through a quantum-mechanical tunneling in which a
> microscopic bubble is formed. Inside this bubble there is
> a new vacuum; the old one is outside it. Once born, this
> bubble will begin to expand rapidly; its envelope, of
> supernuclear density, will acquire a velocity approaching
> the velocity of light, and then our entire Universe will be
> destroyed completely.
>
> When I first had the thought that such a bubble might be
> produced at a particle accelerator in which a particle beam
> was colliding with a target or another beam, shivers ran up
> my spine.
>
> At this point Andrei Dmitrievich interrupted me: 'Such
> theoretical studies must be forbidden'.

I objected that accelerators would go on working regardless of such theoretical work. Furthermore, I said that if the Universe had indeed had an unstable vacuum at some time then it would have been replaced long ago by a stable one, because all possible collisions occurred in the Universe in its infancy.

'But at that time no one was slamming lead nuclei into each other', retorted Sakharov.

In *Sakharov Remembered, A Tribute by Friends and Colleagues*, eds Sidney D. Drell and Sergey P. Kapitza (New York: AIP) 1991, p. 59

I think that the best characterization of Sakharov as scientist can be given by recalling James Gleck's words in *The New York Times* about geniuses, written in an article about the just deceased famous physicist Richard Feynman:

Hans Bethe of Cornell University, paraphrasing the mathematician Mark Kac, said there were two kinds of geniuses. The ordinary kind does great things but lets other scientists feel that they could do the same if only they could work hard enough. The other half performs magic: 'A magician does things that nobody else could ever do and that seem completely unexpected', Dr Bethe said, 'and that's Feynman.'

I would like to paraphrase Bethe: 'That's Sakharov', because Sakharov's ideas and results were 'magic'.

In order to prove that white holes and gorges leading from black holes to white holes can indeed exist in nature, it was necessary to show that these objects are, as physicists would say, stable. This meant that one had to start with analyzing, on the one hand, whether these objects generate processes which destroy them. On the other hand, it was necessary to show that external factors, for example, light rays entering a black hole through the gorge do not destroy the gorge.

The first doubt in the stability of a gorge was formulated by the

British physics theorist Roger Penrose. He pointed to the following feature. Assume that light falls into a black hole through the gorge. Gravitation imparts more and more energy to light quanta. Furthermore, this light energy gets concentrated in a very small volume. Penrose anticipated that gravitation of this compressed energy would destroy the gorge. A number of people were trying to check this conjecture. I had become interested in this problem by the end of the 1970s, even though I was not aware of Penrose's earlier work.

I succeeded in involving three young physicists in working on the subject: Alexi Starobinsky, Zeldovich's postgraduate student, and two of Thorne's students: Yekta Gürsel and Vernon Sandberg. Being together at CalTech in 1978, we vigorously tackled the problem. Time was very short, we regularly worked late into the night.

The result of our work was as conjectured by Penrose, but with a powerful novel feature: the instability that seals the gorge may be triggered even by a single, arbitrarily weak light wave swallowed into the black hole. Our calculations showed that infinitely amplified radiation would create such an intense gravitational field that the gorge would snap shut even before it was fully formed. The entrance to the gorge would be replaced with an impenetrable singularity.

Later the famous Indian physicist Subrahmanyan Chandrasekhar and the American James Hartle developed a complete theory of these processes, and Starobinsky and myself described the quantum processes of particle creation in strong fields inside a black hole, which also result in the formation of an impassable singularity instead of a gorge tunnel. Moreover, white holes also proved to be unstable. It was found that matter from the outside rapidly converts white holes into black holes. This was proved in papers by the American Doug Eardley, the Russian physicist Valeri Frolov and by some others. On the other hand, a joint work by Ya. Zeldovich, A. Starobinsky and myself showed that white holes actively generate matter in quantum processes inside themselves, so that the gravitation of this matter converts them into black holes.

To recapitulate: both white holes and tunnels proved to be extremely unstable and thus cannot exist in nature under 'normal' conditions.

In fact, there is still an option of somehow artificially suppressing the instability and stabilizing the tunnel. This possibility will be discussed in the chapter 'Against the flow'.

Let us return, however, to black holes and try to understand what would happen if an observer dared to set out on a journey into a black hole.

Gravitational forces will pull the spaceship to the region of stronger and stronger forces. At the very beginning of the fall (assume that the ship's engine is turned off) the observer is in the state of weightlessness and feels nothing unpleasant, but this situation changes drastically as the fall progresses. In order to clarify the picture, let us recall tidal forces. These forces arise because the points of a body closer to the gravitational center are pulled more strongly than those farther away from it. The body is therefore stretched out (this stretching happens to the water shell of the Earth, i.e. its oceans, which are attracted by the Moon and thus cause tides).

The tidal stress may be negligible at the start of the fall into the black hole. During the fall, it grows irresistibly. The theory shows that any object falling into a black hole inevitably reaches the region of infinitely high tidal force. Thus any body, any particle is to be torn apart by these forces and to cease to exist. It is impossible to pass through a singularity and avoid destruction.

It was far from easy to prove that the black hole contains a singularity. The decisive step was made by Roger Penrose in 1965. I learnt about his work from Evgeny Lifshitz when Andrei Doroshkevich and myself came to him in order to present our calculations

134

on the formation of a black hole when a 'rippled' spherical body undergoes compression. Lifshitz was keenly interested: 'In fact, you intend to prove that a collapsing not-quite-spherical body produces the same black hole as a perfectly spherical one. However, it is also very important to find what will be the ultimate stage of compression of the body inside the black hole. I have just perused R. Penrose's paper treating this aspect.' And he passed to me the short note of the British mathematician.

In this paper he proved, in exceptionally elegant form, that once a black hole has formed, not letting even light out, then it will contain regions of infinitely strong gravitation, that is, singularities. I still recall the feeling of delight mixed with a certain disappointment. The thing was that I had also tried to find the proof given by Penrose, but failed. The delight was caused by seeing my guess confirmed, and the perfectly understandable disappointment – by the fact that the excellent proof had been found by someone else.

Later, by the end of the 1970s, Penrose and Hawking had proved a number of important theorems about singularities in black holes.

A falling body thus inevitably meets a singularity in a black hole. The reader probably remembers that the radial spatial direction transforms into time in the black hole. The distance from horizon to the black hole center is finite. Hence, the time interval during which a body is allowed to exist within the black hole is finite, in fact, it is extremely short. For instance, in a black hole with a mass of a dozen solar masses it is only one ten-thousandth of a second. For giant black holes of a billion solar masses (such objects very probably exist at the centers of galaxies) this interval reaches several hours. All time lines in a black hole converge to the singularity, and any object will be destroyed in it.

However, if this outcome is unavoidable for anything trapped into a black hole, this means that time also disintegrates in it. The readers may be baffled by this conclusion. 'But what happens *after*

135

it?' you may ask. 'Even the fragments of the destroyed bodies have to exist after the catastrophe, right? Therefore, time will continue ticking, even though such violent processes did take place at this time. Right?'

Wrong. Recall that the properties of time depend on the processes themselves. The theory states that the properties of time in a singularity change to such an extent that its continuous flow breaks down and it splits into quanta. We need to remember at this point that relativity decrees that space and time be treated together, as a unified variety. We should therefore speak of the unified spacetime breaking into quanta.

So far we have no exact theory of this phenomenon. We can only outline the most general features of what should take place. The foremost question is of course: how large or how small are these spacetime quanta? It so happens that this question can be answered even in the absence of an elaborate theory.

Max Planck, who originated the idea of quantization of physical processes, made a conjecture that if a process involves superhigh velocities equal to that of light, plus very strong gravitational fields and quantized properties of matter, then the shortest interval of time can be simply derived from the known values of the speed of light, Newton's gravitational constant and the quantum constant that Planck himself introduced into physics. He calculated that this time interval, known as the Planck time, is unimaginably short. Expressed in seconds, it is given by a fraction with unity for the numerator and unity with 44 naughts (sic!) for the denominator.

Time and space forming an inseparable spacetime, we can also speak of the spatial dimension of these unusual quanta. This dimension (known as Planck's length) in centimeters is given by a fraction with unity for the numerator and unity with 33 naughts (sic!) for the denominator. Planck's dimensions are negligible both in time and in space.

In all likelihood, time intervals shorter than Planck's time are physically impossible. Indeed, we know from quantum physics that there exist the quantum of electric charge and the minimal amount of the energy of light at a fixed frequency of the light wave: the quantum of light.

The existence of the time quantum is not very surprising. The 20th century has made scientific miracles almost a routine matter. Note that this concept of the nature of time stems from the principal inevitability of quantum manifestations in virtually all processes in a singularity.

Once we consider the conditions under which everything is dictated by the quantum properties of matter, time also acquires quantum properties (on a very small scale, though). From this point of view, a constant time flow is in fact an unobservable truly discrete process, similar to the flow of sand in an hourglass, continuous when seen from afar even though this flow consists of discrete sand grains.

In the singularity inside the black hole, time thus breaks into discrete quanta; it appears that as an object approaches the singularity to a distance of Planckian time scale, it becomes meaningless to ask what is going to happen if some more time passes on the clock of the observer falling into the black hole. This time interval is in principle indivisible into smaller parts, just as the energy of a photon cannot be split into parts. The notions of 'before' and 'after' become meaningless and, very likely, the same is true for the question 'What is to happen after the singularity?'

To clarify this remark, we can resort to the following analogy. Recall how an electron in an atom moves along one of its stationary orbits. In the language of classical physics, we say that the electron 'is moving'. In quantum terms, however, 'motion' is not the right word here. It is more correct to say that the electron is in a specific state which is described by the so-called wave function and is independent of time; the wave function makes it possible to calculate the probability for the electron to happen to be in a specific place. It is very likely that the 'flow of time' in a quantum theory of singularity will be described by something similar to the wave or probabilistic function, even though the expression 'the probability for a certain time interval to go by' sounds quite baffling.

Let us recapitulate all this. The properties of time in a singularity are probably changed drastically and acquire quantum features. The river of time breaks into indivisible quanta. It would be wrong to say that the singularity is the time boundary beyond which matter exists outside time. One should rather say that the spacetime forms of the existence of matter change to something so extraordinary that many habitual concepts become virtually meaningless. At our current level of understanding, we can only guess at what laws govern nature in a singularity.

Whatever I was able to say about singularities in black holes was no more than the conclusions of theorists, inferred, of course, from the entire modern physics. This is the frontier of science and many things will be checked and modified. One must not forget that black holes, which inevitably contain singularities that block the flow of conventional continuous time, are real objects in the Universe. Astrophysicists have already discovered, with a high degree of certainty, several such objects. What has been found is a kind of sink of the river of time: maelstroms that never let anything out.

8

Energy extracted from black holes

Our story of holes in space and time would not be complete if we failed to mention their wonderful property of continuously releasing energy. This feature is one of the manifestations of the as yet undeciphered relationship between time and energy. This relationship manifests itself clearly when quantum properties of matter begin to dominate.

However, I should start very briefly with empty space and its quantum properties.

According to current notions, the vacuum is not absolute emptiness, the 'perfect nothingness'. It is a sea of so-called virtual particles and antiparticles which do not emerge as real particles. However, the vacuum is the place where pairs of virtual particles and antiparticles are constantly created for a very short moment, only to disappear immediately. They cannot transform into real particles because this would mean the creation of real energy from emptiness. The so-called uncertainty relation of quantum physics allows these particles to appear for a fleeting moment; this relation states that the product of the lifetime of a pair of virtual particles and their energy is of the order of Planck's constant. Real particles can always be removed from a volume while virtual particles cannot be removed – in principle.

Such are the properties of the vacuum. If some strong field is applied to the vacuum, then some virtual particles may 'pick up' sufficient energy in this field to become real; they extract the energy for that from the external field. This is the mechanism for creating real particles from the vacuum at the expense of the energy of the strong field.

The facts have been well known for a considerable time; for example, such charged particles as electrons and positrons are indeed created from the vacuum in strong electric fields.

Let us turn now to black holes. In 1977 Yakov Zeldovich and Alexi Starobinsky analyzed processes in the vacuum in the neighborhood of a rotating black hole. The problem is that when a rotating mass is collapsing and a black hole is formed, there is not only the gravitational field that pulls all bodies towards the center but also a field which forces all moving objects to rotate around the black hole; that is, a swirl-like gravitational field is formed. Such black holes are known as rotating black holes.

Zeldovich and Starobinsky were able to show that radiation quanta are created in the vicinity of such black holes at the expense of the energy of the rotational gravitational field. As a result, the energy of the black hole gradually transforms into energy of radiation. This is a very slow process. For example, for a black hole of ten solar masses, rotating at maximum possible speed, only several hundredths of one erg of energy has been radiated away during the entire lifetime of our Galaxy (about ten billion years). This is an absolutely negligible amount.

In the autumn of 1973 Zeldovich and Starobinsky reported their calculations to Stephen Hawking during his visit to Moscow. After his return to Cambridge, Hawking started checking the conclusions using his mathematical methods. He later recalled:

> However, when I did the calculations, I found, to my surprise and annoyance, that even nonrotating black holes should apparently create and emit particles at a steady rate. At first I thought that this emission indicated that one of the approximations I had used was not valid... However, the more I thought about it, the more it seemed that the approximations really ought to hold... Since then the calculations have been repeated in a number of different forms by other people. They all confirm that a black hole ought to emit particles and radiation as if it were a hot body with a temperature that depends only on the black hole's mass: the higher the mass, the lower the temperature.

This was a spectacular discovery.

I will try to clarify, at least very approximately, how radiation is emitted. The essential thing about the process is that it is of quantum nature. Virtual particles are created in the vacuum at some distance from one another. In the case of the gravitational field of the black hole, one particle can be created outside the horizon, the other inside the horizon. The particle born outside the horizon can fly away into space while the other will keep falling into the black hole and will never reach the remote observer. The two will never be able to merge and disappear as happens with virtual particles in the usual vacuum. Thus a flux of particles away from the black hole is formed in space. Some energy of the black hole is therefore expended and the size of the black hole decreases. Stephen Hawking proved that the black hole radiates energy as if its surface is heated to a certain temperature.

I must immediately emphasize that the temperature of stellar-mass black holes is absolutely negligible. For example, in a black hole of ten solar masses, the temperature is only one ten-millionth of one kelvin. The greater the mass, the lower the black hole temperature, so that the temperature of supermassive black holes is absolutely negligible. On the opposite, the lower the black hole mass, the higher its temperature and the faster goes the process of transformation of black hole mass to radiation. As I have already mentioned, stellar-mass black holes emit negligible amounts of mass. Under natural conditions, they absorb much more energy in incident radiation and rarefied matter. However, a sufficiently small black hole can emit energy at a considerable pace, and thus must be treated seriously as an energy source. For example, a black hole with a mass of a billion tons (the mass of a modest mountain) will emit one hundred million billion ergs per second over ten billion years. Its temperature in the process will be about one hundred billion kelvin. Note that this is nearly ten thousand times higher than the temperature in the depths of the Sun. The size of this black hole is supermicroscopic: about the diameter of an atomic nucleus.

The extremely slow process of energy loss by a black hole to quantum radiation is known as quantum evaporation; however, the radiation of energy by low-mass black holes cannot be referred to as evaporation: this is a very substantial glow. In the course of this glow, the mass of such stars diminishes at constantly increasing rate. When the mass drops to one million tons, the temperature reaches one hundred billion kelvin. The process of emission turns into explosion. The last thousand tons explode in one tenth of a second, releasing energy equal to that produced by one million one-megaton hydrogen bombs. The quantum energy release by low-mass black holes is thus a highly efficient process. However, can such black holes be formed at all?

I have already emphasized that artificial manufacture of a black hole is utterly unrealistic, at least in the foreseeable future of science. Could nature have produced such objects?

We will see later in the book that this question is answered in the affirmative. Mini black holes could have been created at the early stages of the expansion of the Universe. Why aren't they being created in today's Universe and why would it be extremely difficult to produce them in a laboratory in the remote foreseeable future, let alone in contemporary ones?

The point is that this requires compressing matter to very, very high density. To turn the Sun into a black hole, its matter must be compressed to nuclear density, while to turn the Earth into a black hole, its matter would require compression to a density even one hundred billion times higher.

This tremendous compression calls for huge forces. In very massive stars, these forces are exerted by their gravitation. However, gravitation is obviously insufficient for small masses, so that very high external pressure is required. Such colossal forces cannot be found in nature, nor in man-made laboratories (it is not clear whether it will ever be possible to create such equipment).

If, however, we turn to the past history of the Universe (we will discuss it later), we easily notice that the conditions at the very beginning of the expansion, about 15 billion years ago, were favorable for the creation of small black holes. Indeed, the entire matter was in a state of tremendously high density and no additional compression was needed. Actually, this matter was expanding at a very high speed. Therefore, a black hole could form if either the velocity of expansion in a small volume was somewhat lower, or the amount of matter was somewhat greater than in neighboring volumes of the same size. Gravitational forces could then slow down the expansion in this volume and some time later turn it into compression, producing a mini black hole. Zeldovich and myself in 1966 and Hawking in 1971 pointed to this possibility. In the scientific literature, the theoretical discovery of the possible formation of such primordial black holes, especially black holes with masses much less than the mass of the Sun or other stars, is often ascribed to Stephen Hawking. I believe that the reason for this is as follows. Even though Yakov Zeldovich and myself clearly understood this possibility and often mentioned it years before Hawking's paper was published, we never showed with sufficient clarity and decisiveness that black holes with masses much smaller than stellar masses could be formed in this way. I believe that this clarity is mandatory if authors wish to be correctly understood, otherwise your pronouncements are likely to remain 'a thing unto itself', comprehensible only to you, or, as Zeldovich used to say, remain a V-sign given behind your back. (*Translator's remark*: This seems to be the nearest to the typically Russian positioning of fingers and the message it carries that is prominent in the image used by Zeldovich.)

We thus conclude that tiny black holes could exist in the Universe at the early stage, and that their masses could be much smaller than stellar masses. What was the fate of these objects? It depended on their masses. Small black holes began to emit via the quantum mechanism. Calculations show that all black holes whose initial mass was below one billion tons had 'evaporated' completely by our time. Heavier black holes survived till our days. Can these be detected by astronomical means, assuming they do exist in the Universe?

The most direct approach to finding them would be to look for their emission of hard quanta. Observing such quanta coming from the cosmos could help in identifying primordial black holes. None have been detected so far. We can only conclude that the number of black holes with a mass around a billion tons does not exceed a thousand per cubic light year. If their number was higher, their aggregate emission would be detectable. The number 'one thousand' looks impressive but do not forget that their total mass is negligible in comparison with that of stars.

Only future measurements will show if there are mini black holes in the Universe.

It must be clear from our story as it unfolds that the Universe is likely to have black holes of stellar origin, both supermassive ones at centers of galaxies and mini black holes of the early Universe. All these black holes may serve as sources of energy in the future. In the case of massive black holes, it is possible to make use of their gigantic gravitational energy. I will not go into a description of how this may be done – in principle. This would lead us too far astray from the main topic. As for the mini black holes, they continuously emit energy.

There could be different ways of utilizing this energy. For example, we can imagine a necessary number of mini black holes revolving around the Earth and emitting their quantum radiation. However, how do we place such a black hole on an orbit around the

145

Earth? In fact, how do we transport a black hole? This is not a conventional body, it has no material surface, it cannot be hooked by a cable and towed to the right place. You cannot fix a jet engine to it and start going. Finally, you cannot lock it into a container. Indeed, although having the mass of a mountain, it has the size of an atomic nucleus. It would freely pass through any obstacle, cut through the entire globe.

Is there a way of forcing a mini black hole to move in a desired direction, of making it increase or decrease the speed of this motion? Let us fantasize about it. How could we make this black hole obey a command?

Our first approach, of course, is the gravitational field. A black hole obeys gravitation in exactly the same manner as any other physical matter. It falls in this field at the same acceleration of free fall as other bodies, and bends its trajectory as they do. Clearly, therefore, the simplest method of urging it to move in a prescribed direction is to apply a gravitational field.

For example, we can do the following (see figure 8.1). Let us guide to the vicinity of the mini black holes a sufficiently massive body, say, an asteroid more massive than this black hole. We can do this using jet engines mounted on the asteroid. The black hole will start falling in the gravitational field of the asteroid, towards its center of mass. We can wait a bit while the black hole gains sufficient velocity in the right direction, after which the asteroid can be taken out of the way and the black hole will continue to travel by inertia at the acquired velocity.

Of course, with an asteroid of a modest mass and realistic size, the acceleration created by it cannot be high. The velocity gained by the black hole cannot be high either. For example, an asteroid of a radius one hundredth that of the Earth could accelerate a mini black hole to a velocity of about one hundred meters per second.

Fig. 8.1.

The method can be improved, however. The jet engines on the asteroid can be programmed to give the asteroid an acceleration away from the black hole, equal to the acceleration with which the black hole falls onto the asteroid. In this case the velocity of the black hole–asteroid system can grow slowly but steadily.

The black hole can be similarly decelerated by bringing an asteroid from the opposite direction, or its direction of motion can be changed. If the black hole is already on an orbit around the Earth, then bringing massive bodies to the right side of the black hole we could correct its trajectory by the gravitational field of these bodies.

Here is another, similar method of transporting a black hole. Let us guide a massive asteroid to the black hole and program its maneuver in such a way that having approached it, the asteroid would thereby force the black hole to move on a circular orbit

147

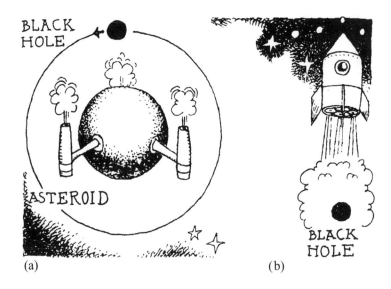

Fig. 8.2.

around itself (see figure 8.2(a)). After this the asteroid can be slowly accelerated by the jet engines that it carries. If the acceleration phase is sufficiently gentle, the black hole should follow the asteroid always staying on the orbit around it. This process is possible if the asteroid's acceleration is smaller than the acceleration of free fall of the black hole to the asteroid.

The methods outlined above all require the use of very massive bodies. Can we avoid using them?

It is found that we can. One such method is shown in figure 8.2b. A rocket with its jet engine working 'hangs' above a mini black hole. The jet of high-velocity hot gas mostly passes by the black hole but a small part falls into it. As a result, the entire 'rocket plus black hole' system gains velocity towards the nose of the rocket and gradually accelerates. The nearer the rocket to the black hole, the

Fig. 8.3.

harder must the jet engines work to save the rocket from falling; hence, the faster the system accelerates.

Figure 8.3 shows another method of imparting velocity to a black hole, this time without gravitational fields or rockets. It is possible to irradiate the black hole by a directed flux of radiation which will be absorbed by the black hole; the black hole gains the momentum carried by the radiation flux and starts moving. We might say that the black hole is made to move by pressure of radiation. Isn't it an extraordinary situation when the pressure of radiation acts on emptiness, or rather, on a blob of gravitation (which a black hole is).

Let us put a stop to this run of fantasy (so far fantasy). The main purpose of this chapter was to show that holes in space and time are not eternal. Hawking's radiation slowly evaporates them.

It is not quite clear yet what significance this has for the theory

149

of time. It is clear, nevertheless, that black holes, acting as sort of 'sinks' for the river of time, slowly heal up. The Russian academician M. Markov believes, for example, that the ultimate residue of a black hole must be an elementary black hole: a particle with a mass of one hundred thousandth of one gram.

This and many other aspects of black hole physics are the subject of intense research for theorists.

9

Towards the sources of
the river of time

We have thus far discovered that time can flow very differently. The river of time has estuaries and sinks. However, does it have a source?

Now that it is clear that the properties of time are a function of the physical processes that go on in nature, this question no longer seems to be so absurd. Philosophers pondered this problem for quite a long time. However, the striking successes of Newtonian mechanics and, as a result, the universally accepted Newtonian concept of eternal and unchanged time accustomed them to thinking that the source of time was in the infinite past.

Time was thought of as a uniform river or a never changing road which stretched from the past to the future. In fact, scientists had to face the problem of the beginning of time again, and quite dramatically, in the 20th century. This happened after the discovery of the expansion of the Universe. A detailed description of this achievement is given in the book *Edwin Hubble: the Discoverer of the Big Bang Universe*, that A. Sharov and I wrote in 1989. (An expanded version in English translation was published in 1993 by Cambridge University Press.) Here I will only trace the main points on this road.

It all began at the end of the 19th century. A rich American astronomer Percival Lowell had a private observatory built for him in the Arizona desert. He dared to do this because he was very keen on the observations of the Italian astronomer Giovanni Schiaparelli of the mysterious lines on the Martian surface, which he interpreted as canals. Lowell was also interested in the origin of the Solar System. He strongly believed that some nebulae observed in the sky were planetary systems in the process of formation. Among such objects he counted the nebula in the Andromeda constellation. We know very well now that the Andromeda nebula, shaped as a spiral curling towards its center, is one of the stellar systems closest to us resembling our Milky Way galaxy. However, nothing was known in

Lowell's time about the tremendous distances separating the nebulae from us.

Lowell suggested to a young astronomer, Vesto Melvin Slipher,
whom he had recently hired to work in his observatory, to take
up spectral studies of the Andromeda nebula. It was a challenging task. The nebula's brightness is very low: it is barely visible to
the unaided eye. The sensitivity of the photographic plates, used
to record the spectrum, was rather low at that time and the telescope was quite modest by our standards. It was a refractor (using
lenses) with the objective lens diameter of 60 cm. For comparison, one of the largest reflector telescopes in the world in North
Caucasus is 6 m in diameter and collects a hundred times more
light.

On the night of 17 September 1912 Slipher obtained the spectrum of the Andromeda nebula after a seven-hour exposure, and
was able for the first time to measure its velocity, using the Doppler
effect. The astronomer could not believe his eyes. The velocity was
huge: the nebula was speeding towards us at 300 kilometers per
second. Slipher made several more photographs of the spectrum
which confirmed this result, and only then published the paper.
The velocity of the Andromeda nebula was ten times the ordinary
stellar velocity. Slipher realized that he had come across something
quite extraordinary. In his paper he wrote: 'Thus the extension of
the work to other objects promises results of fundamental importance'.

Slipher outlined the program and started implementing it.

It was a work of exceptional complexity. Other nebulae are so dim
that exposures of dozens of hours were required. Slipher repeated
his sessions night after night. After two years of toil he was able
to measure the velocities of 15 nebulae; he kept accumulating
the observational material. The velocities were staggeringly high.
Almost all nebulae, except for the Andromeda and several others

visible in almost the same area of the skies, were found to be moving away from us. The maximum recession velocity was 1100 kilometers per second.

In 1917, Vesto Slipher made a summary report on his painstaking effort. He concluded that, first, nebulae are not seeds of new planetary systems. Second, he was able to formulate his main result on the basis of the velocities of 25 nebulae measured by that time: *taking into account the sign, the average velocity is positive; it indicates that nebulae recede at a velocity of about 500 km/s. This could mean that spiral nebulae go away from us but their distribution on the sky does not agree with this interpretation since they tend to form clusters.*

This was the first, tentative and qualified, formulation of a suspicion that the world of nebulae is expanding. Nobody was quite sure at the time that nebulae are stellar systems, or galaxies, similar to our Milky Way galaxy.

Several years later another American astronomer Edwin Hubble was able to prove that nebulae consist of stars, and measured the distances to them. He found that the distances are tremendously large and that nebulae are in fact huge stellar galaxies.

The next truly great discovery made by Edwin Hubble was the law describing how galaxies move away. Looking for the correspondence between galactic recession velocities and the distances to them, Hubble found in 1929 that the recession velocity is directly proportional to distance. This is the great law bearing his name. Of course, galaxies recede not only from us (from our Galaxy) but also from one another, so that the whole Universe expands.

Hubble's discovery followed in the footsteps of theoretical work which attempted to describe the structure of the Universe in terms of new physical theories.

The most important of these theories was Einstein's general relativity, which related gravitation to the geometrical properties of space and to the slowdown of time flow in strong gravitational fields.

Soon after Einstein completed his general relativity theory, he constructed on its basis a theoretical model of the Universe. No observational evidence of systematic motion of remote worlds existed at that time, and Einstein assumed that on the largest scale matter is at rest once we average the local motion of individual objects whose velocities are relatively small (tens of kilometers per second). However, this fact contradicted the theory of gravitation!

Indeed, the only significant forces acting on the scale of the Universe are the forces of universal gravitation. Therefore, if we assume that at some moment the enormous masses of the Universe are at rest relative to one another, at the next moment they will start moving towards one another. This was the conclusion already made by Isaac Newton. The same result obtains in Einstein's new theory. All matter on the vast scale of the Universe must start contracting. Immense star worlds – galaxies – can be treated as 'particles' of this matter; hence, galaxies must tend to come closer.

However, Einstein did not believe in this scenario; he knew that astronomical observations of the time did not point to anything of the sort (this was long before Hubble discovered the expansion of the Universe). Einstein assumed, therefore, that in addition to the attractive gravitational forces, nature must also have hypothetical repulsive forces that science knew nothing about. Correspondingly, Einstein introduced into his equations the so-called cosmological term describing the cosmic repulsive forces. These forces were set to be extremely weak and to manifest themselves only over very large distances on the scale of the Universe, being absolutely insignificant on the Earth or even within the Solar System.

These invented repulsive forces were meant by Einstein to balance out the gravitational forces and make the Universe static.

For this static state to exist at all times, it is not sufficient for the forces to cancel each other out; obviously, the equilibrium must restore itself after each small accidental low-scale local motion of matter; in physicists' parlance, the equilibrium must be stable.

This picture looked a bit too speculative. However, only a handful of outstanding theoreticians thoroughly understood the complicated theory developed by Einstein. Several, and among them the outstanding British theorist Arthur Eddington, began to think about these problems.

Answers to these theoretical questions came in the mid-1920s from Soviet Russia.

Aleksander Friedmann, a mathematician from Petrograd (the city that we now know as St Petersburg), solved the equations of Einstein's new theory for the motion of mass on the scale of the Universe. The conclusions from his solution were extremely important.

Friedmann showed that regardless of whether the hypothetical repulsive forces postulated by Einstein exist or there are only the universal gravitational forces, the Universe has to be nonstatic.

Indeed, the repulsive and attractive forces can equilibrate each other only at a very specific density of matter in the Universe. Even a minute deviation from this state makes one of the two forces greater than the other and the Universe must start expanding or collapsing. The initial conditions should thus decide whether the result is expansion or contraction. Even without the hypothetical repulsive forces, the Universe will fly apart if initially its masses moved away from one another (the initial velocities were caused by some processes in the past that we shall discuss later in the book); the mutual gravitational attraction will slow down the outward motion. Whether the Universe will expand or contract thus depends on the processes that dictated the initial velocities of the masses.

The equations derived by Friedmann describe not only the

dynamics of motion of masses in the Universe but also the geometric properties of space; mathematicians say that they describe the curvature of space, which changes with time as the Universe expands.

Friedmann's conclusions were at first flatly rejected by Einstein but after clarifications sent to him by Friedmann with a physicist Yu. Krutkov, Einstein completely accepted Friedmann's results.

What happened next is hard to comprehend. Even though Friedmann's papers were published by the widely known German journal which also published Einstein's admission of his initial misjudgment, these papers completely missed the attention of not only astronomers (this was not surprising: astronomers find it quite hard to follow the latest news in theoretical physics; alas, this happens to astronomers even now) but also of physics theorists. The latter is not easy to explain. Nevertheless, I will offer, a few paragraphs later, some hypotheses that may help to clarify how this became possible.

In the West, the theoretical aspects evolved largely independently.

The Dutch physicist Willem de Sitter had already analyzed, in 1917, a theoretical model of the Universe almost devoid of all gravitating matter; that is, practically empty but dominated by Einstein's forces of cosmic repulsion. Six years later, the German mathematician Herman Weyl noted that if galaxies are inserted into this Universe at very low density, so that their attractive gravitation is negligible in comparison with repulsion, they will gain velocities proportional to the distances between them (provided the intergalactic distances are not too large).

Five years later another theoretician, H. Robertson arrived at the same conclusion. He compared the distances to galaxies, calculated from Hubble's observations, with the velocities found by Slipher. Robertson noticed an approximate proportionality of velocities and distances. Soon this law was very definitely established by Edwin Hubble.

About a year before Robertson, Arthur Eddington's student Georges Lemaître re-derived the equations obtained five years before that by Aleksander Friedmann. Unlike Friedmann, however, he paid attention to the astronomical observations of remote galaxies, which could test the validity of his theory.

When Sharov and I were describing all the dramatic bends on the road to discovering the expansion of the Universe, collecting documents, re-reading publications, talking to our Western colleagues, we were desperately trying to understand why the significance of Friedmann's work was not appreciated by contemporary scientists.

Probably, one of the factors that contributed to this neglect was that Friedmann failed to mention the observational testing of his theory while other theoretical papers discussed this aspect, which was closer and more understandable to observational astronomers. Consequently, they paid more attention to papers discussing the observational tests.

At the beginning, neither Hubble nor other people involved in the first discussions of his discovery knew or remembered even the early publications of theorists in the West, let alone Friedmann's work. In all likelihood, de Sitter's model which predicted the recession of galaxies in the almost empty Universe and Einstein's static model were the only results taken into account at the time.

Why was it that both Friedmann's work and that of a number of theorists in the West that predicted a non-stationary Universe remained either unknown to astronomers – for quite a long time – or failed to arouse their interest? We can offer several possible reasons for the strange situation in which, for some time, a theoretical prediction of a most important natural phenomenon failed to draw the attention of precisely those people who were able to test this prediction.

One of the likely reasons seems to be that new cosmological models were based on general relativity which is very complicated not only in its mathematical tools but, most importantly, also in new concepts of space, time and interpretation of the gravitational interaction. At that time not only observational astronomers but even theoretical physicists were slow in grasping new ideas and in clearly understanding them, and also in the desire to make use of them in specific research projects.

The first reason, therefore, seems to be the gap between theorists and observers. Another reason was psychological. It stemmed from the extraordinary corollaries from the theory which stated, for example, that the space of the Universe may be closed, or that our world may have evolved throughout its past. Practical astronomers who used their new powerful telescopes to penetrate deeper and deeper into the cosmos, found it difficult to accept such conclusions which drastically re-wrote their general picture of the Universe.

However, let us forget these speculations and return to the situation that existed in astronomy after these discoveries had been made.

In the 1920s both theorists and observers had thus established that we live in an expanding Universe which 'exploded' at some moment in the past.

This discovery overturned the notion of the Universe as something grandiose and – on the average – unchanging, containing the eternal re-circulation of matter.

There can be no doubt that a discovery of this scale should be decisive for the understanding of the nature of time in the Universe.

Almost seventy years have passed since Hubble's discovery. The research into the phenomenon he opened for us still continues. In the words of the famous Italian philosopher Giordano Bruno, 'The will that strives for understanding is never satisfied with the job completed.' The equipment at the disposal of today's observers – their telescopes and instruments – greatly surpasses anything in the past. Astronomers now scrutinize galaxies that are separated from us by ten billion light years and receding from us at almost the speed of light. For example, the velocity of recession of the farthest known quasar (bright galactic core) is so grandiose that the wavelengths of light emitted by this source are increased by the Doppler effect fivefold.

In our time the theory of the expanding Universe is excellently supported by observational data. Also, it offers conclusions of principal importance. One of these conclusions concerns the curvature of three-dimensional space.

Friedmann's theory states that if there is a sufficient amount of matter in the Universe, so that its average density is greater than a certain critical value, then the curvature of space due to the gravitation of this matter resembles the curvature of a sphere. This difference stems from the fact that the spherical surface is two-dimensional while space is three-dimensional. The spherical surface curves and closes on itself; its surface is finite. Likewise, curved three-dimensional space closes on itself. The world is closed.

Of course, it would be very difficult to build a mental picture of such a closed world. The famous French philosopher Blaise Pascal wrote: 'Imagination will tire comprehending sooner than nature will tire supplying'. Science has in fact taught us to accept that phenomena exist which can hardly be explained by a visually clear image. If the average density of matter in the Universe is equal to or below the critical value, space is infinite. This is fine; but is our actual Universe finite or infinite?

We do not have the final answer to this question. The critical average density of matter is calculated from the rate of expansion of the Universe. It equals roughly five solar masses in a cubic box with each edge a thousand light years long. It is extremely difficult to measure the actual matter density in space for a comparison with the critical value. The reason is that considerable amounts of very dim or even completely invisible matter are very likely to float around galaxies and in the space between them. Astrophysicists refer to it as 'hidden' or 'dark' matter. To detect this matter or take it into account in calculations is an especially difficult task. The total mass of the visible stars, planets and gas gives a density about one thirtieth of the critical level. The 'hidden mass' is about thirty times greater than the visible mass. It is therefore unknown if the actual density of matter in the Universe is greater than the average density or not, and hence if our world is closed or infinite.

The following is, however, clear. Even if the Universe is closed, its size is immense. It is much larger than the distance to the most remote observable galaxies, that is, much greater than ten billion light years.

Another corollary of the theory of the expanding Universe is especially important for the problem of the sources of the river of time.

Since the Universe expands, galaxies were closer to one another in the past than they are now.

Still earlier, though, no galaxies or any separate heavenly bodies could exist. In that distant epoch, there was only dense expanding matter which only much later split into fragments from which systems of heavenly bodies later formed.

Even earlier still, in the very distant past, there was a moment when, according to the theory, matter density was formally infinitely high. This was the moment when expansion started. The state of the Universe at this moment is known as a 'singularity'.

How long ago did this expansion start? Calculations based on the observable velocity of recession of the galaxies show that this outward motion started about 15 billion years ago. What was the nature of this event? What was there 'before'? What were the properties of space and time close to the singularity? These are the great mysteries of our Universe.

20th century science has made impressive contributions to solving the singularity problem. Friedmann's theory describes how expansion is affected by the gravitational interactions. Galaxies recede from one another by inertia while the mutual gravitational attraction gradually decelerates their motion and slows down the expansion of the Universe. A comparison of the conclusions of the theory with observational data shows that the expansion started about 15 billion years ago. The theory does not answer, however, what triggered the expansion, or what imparted to the matter from which galaxies were later formed the initial expansion velocities.

A discovery made in 1965 was expected to provide answers to all these questions. A very weak electromagnetic emission was discovered, at a temperature of about three degrees on the Kelvin scale, which uniformly fills the entire Universe. Electromagnetic radiation was present in the Universe from the very beginning of the expansion; as we already know, Shklovsky gave it the name 'relic radiation'. It has cooled down to its present temperature in the course of the expansion, while in the past its temperature, and that of matter, reached extremely high values. The Universe was hot and the pressure of matter, which was distributed in space almost uniformly, was tremendous.

At first glance, the high pressure seems to be of primary importance for an explanation of the explosive expansion of the Universe. Remember what you know about a bomb exploding. The explosion heats and evaporates the charge, the pressure of hot gases rapidly expands the matter of the charge: the bomb explodes. It may seem that the Universe began its expansion in the same way. One may decide that very high temperature and colossal pressure were the causes of the expansion. This conclusion would be very wrong. There is an essential difference between the two phenomena.

The explosion of a conventional explosive produces a pressure drop: from the very high pressure inside the hot gases to the relatively small atmospheric pressure outside. This pressure drop produces the force that throws the matter apart, not the high pressure as such. If the pressure outside was the same as that inside the gas, matter would not fly apart. Furthermore, the density of the expanding gas is nonuniform in the explosion: it is maximal at the center and drops off away from it. As the matter flies out, the pressure drop caused by the drops in density and temperature creates a force that propels the expanding gas.

The starting moment of the expansion of the Universe is very different from the above picture. Its matter was uniformly distributed in space before individual more or less compact bodies began to form. Temperature was indeed very high but the same in the entire space. There were no density and pressure drops, so no force could arise to cause the expansion. Therefore, high density of hot gas was not something that triggered the expansion. What then was the 'primary push' that gave initial velocities to matter?

To understand this, we need to return to 'the very beginning'. This requires that we learn about the properties of matter at extremely high densities and temperatures.

10

Journey to unusual depths

In our journey towards the sources of the river of time, we have to face the fact that the closer to the singularity, the higher the temperature of the Universe, and hence the higher the energy of particles of matter. What processes are we to envisage in this world of enormous energies? To sort out this side of the story, we leave cosmology alone for a while and venture into the world of the infinitely small: into the world of modern elementary particle physics.

This trip will have to be rather short, sufficient only to make acquaintance with the facts that are important for understanding the processes occurring in the early Universe.

A true revolution has taken place in modern elementary particle physics during the last quarter of a century. It is now clear that all the elementary particles of which matter consists, such as the proton and the neutron, are not 'elementary building blocks' of nature at all: they are complex systems composed of even smaller elementary objects called quarks. The existence of whole classes of new particles with quite extraordinary properties has been established. But the most important milestone was probably the discovery of the spectacular unity of various fundamental forces which until relatively recently were regarded as very dissimilar. This unity manifests itself at very high energies and is therefore especially significant for the understanding of how the Universe began to expand.

It was not the first time that physicists had realized that forces so unlike one another are in fact different manifestations of one, more general entity. The same happened with the electric and magnetic interactions. People had been familiar with the manifestations of these forces since time immemorial and assumed that magnets do not affect electric charges, and vice versa. However, the experiments of André Marie Ampère, Michael Faraday and others showed that moving charges create magnetic fields, while the motion of a magnet generates electric forces. Fifty years later the electromagnetic theory of James Clerk Maxwell unified these apparently dissimilar interactions into a unified entity: the electromagnetic field. It had just been discovered that electromagnetism is one phenomenon, and it only 'splits' into electricity and magnetism in special conditions, when the fields do not change in time.

Soon after creating general relativity, Albert Einstein began a titanic effort, trying to unify electromagnetism and relativity: the two types of interaction that were known at the time. He kept working on this program throughout the rest of his life. However, science was not ready at that time to successfully complete the effort, not even to appreciate the grandiose scale and significance of these efforts. A considerable number of physicists regarded Einstein's attempts with utmost skepticism. For example, the famous physics theorist Wolfgang Pauli used to say metaphorically: 'Man cannot join what God deemed to make separate'. When attempts to fuse other forces of nature were made later, they often met with similar skepticism.

In the spring of 1988, in Trieste, I asked a renowned Pakistani physicist Abdus Salam, director of the International Research Center there, about the earliest attempts at creating theories unifying different forces. Salam answered that almost nobody believed in them thirty years ago and remembered a letter that Pauli wrote to him in 1957. The letter went roughly like this: 'I am now reading your paper (on the shores of the Zurich lake, under bright sun). The title, *The universal Fermi interaction*, surprised me greatly, here is why. For some time already, I follow this rule: if a theoretician says "universal", he describes sheer nonsense.'

Many decades have passed since Einstein's first efforts; the situation in physics has changed drastically since then. We now know four types of physical interaction: the gravitational, the weak, the electromagnetic and the strong.

We have mostly spoken so far about the gravitational interaction that controls the motion of heavily bodies; it can be safely ignored in the world of elementary particles. Several explanations must be given now about the other three types of interaction.

An example of a process going through the weak interaction is the decay of a neutron into a proton, an electron and an antineutrino. We find an essential difference between this interaction and the manifestations of the gravitational interaction as discussed earlier. In the context of the previously mentioned slow motion, only the state of motion of particles is changing; in contrast to this, the weak interaction changes the nature of particles: a neutron is replaced with a proton, an electron and an antineutrino.

Strong interactions produce various nuclear reactions (e.g. thermonuclear, or fusion reactions) and also the forces that bind neutrons and protons into atomic nuclei.

We are familiar with electric and magnetic forces through experiments in high school, so no comment is needed here.

All processes in the Universe are the results of these four types of interaction. What actually happens in these interactions? What is the most important thing about them? Particles interact via the exchange of other particles – carriers of interactions. Each of the four types of interaction has its own carriers.

The electromagnetic interaction is mediated by the photon, its 'carrier', and the gravitational interaction is mediated by the graviton. These two always move at the speed of light and have no mass of their own. Physicists say that their mass (another expression is 'their rest mass') is zero.

The weak interaction also has its 'carriers'. These are particles that physicists call 'vector bosons' (I will not go into explanations of why this name was chosen). The essential difference between these and the photon and the graviton is that they are extremely massive: about a hundred times heavier than the proton. Because the carriers are so heavy, the weak interaction is only possible at extremely short distances. This distance is about a thousand times shorter than the size of an atomic nucleus. Recall that a nucleus is about a hundred thousand times smaller than the atom. Why does the weak interaction act only over such a short range? The reason is this. To emit a heavy carrier particle, an interacting particle needs to spend a great deal of energy, and there is nowhere to borrow this energy from! However, the world of elementary particles obeys the uncertainty relation. I have already mentioned this relation in the chapter 'Energy extracted from black holes'. According to this relation, a particle or a system may gain energy as if from 'nowhere' but for a very short duration. The higher this energy, the shorter the time interval during which it can be 'borrowed'. Then this 'borrowed' energy has to be paid back, otherwise it would constitute a contradiction with one of the fundamental laws of nature: the law of conservation of energy. In this way, a particle can create the 'vector boson' carrier by borrowing energy 'from nowhere' only during one millionth of a billionth of a billionth of a second. This is the maximum interval between the emission of a weak interaction carrier and its reabsorption. No wonder, therefore, that even moving at the speed of light, the carrier can only cover during this time no more than a thousandth of the diameter of the atomic nucleus. This gives us the radius of action of the weak nuclear forces.

The example discussed above shows a very specific relationship between energy and time in the world of elementary particles. Indeed, the amount of energy borrowed from 'nowhere' and the interval of time after which the 'loan' must be returned are related by a strict mathematical formula: the higher the energy, the shorter the time interval. Note that any substantial amount of energy can be 'borrowed' for only an extremely minute duration.

Another manifestation of the relationship between time and energy must be mentioned here: the law of conservation of energy, discovered by physicists a long time ago.

That energy cannot 'appear from nowhere' for prolonged periods (not for the infinitesimal intervals discussed above) was established after numerous failed attempts over several centuries to invent a *perpetuum mobile* – the perpetual engine. The law of conservation of energy was formulated in 1842 by the German medical doctor Julius Mayer. It is a curious fact that Mayer came to this conclusion after sailing to the Indonesian island of Java as a ship's doctor. Observations of the venous blood of sailors led him to a hypothesis that mechanical work and heat may convert into each other. In 1842 Mayer published a paper 'Notes regarding the forces in lifeless matter', in which he formulated his law of conservation and transformation of energy. Several years later, this law was rediscovered by James Joule and Hermann Helmholtz. Mayer's work remained unrecognized. He tried to defend his priority of the discovery, which led to serious nervous disorder. In 1862, Rudolf Clausius and John Tyndall noticed his work and his priority has been recognized.

The law of conservation of energy says that the energy of a system that is isolated and does not interact with other systems cannot change. It does not change with time.

The profound nature of this fundamental property was revealed in 1918 by the German mathematician Emmi Noether. She was able to show that energy is conserved because time is uniform. According to Newton's physics, all moments of time are equivalent. For this reason, as shown by Noether in all mathematical rigor, energy remains the same at all moments of time. This was a very novel approach to physical laws, based on symmetry properties, in this case time symmetry. It was also found that other physical properties – momentum of a body and its angular momentum – are conserved in time also owing to symmetry properties, this time space symmetry.

This was the first discovery of the profound relation between physical properties and the symmetries of space and time. We will see later in the book that the application of the symmetry concept is one of the most important ideas in modern physics.

Let us turn now to the strong interaction. Its carriers are *gluons*. They are emitted and absorbed by quarks of which (as was mentioned at the beginning of this chapter) consist neutrons and protons, as well as some other particles. Like photons, gluons have zero rest mass. In the case of electromagnetic interactions, the emission and absorption of carriers are due to the electric charge of particles. In the case of strong interactions, the emission and absorption of gluons is also due to special charges possessed by quarks. However, these charges can be of three types which were given the names *red*, *yellow* and *blue*. The strong interaction is sometimes called the color force. A quark has one of these three 'colors'. Of course, these labels have no relation at all to the ordinary colors of things.

Another property distinguishing the strong interaction from electromagnetism is that gluons also carry color charges, that is, are color-charged. We find nothing similar to this in the case of electromagnetic forces. Their carriers – photons – are electrically neutral and possess zero electric charge.

It may seem that we could make a stop here on our journey to the microscopic world, in our familiarization with the smallest (known today) particles of matter. Actually, though, the quite reliably established facts outlined above are only an introduction to acquaintance with the truly awesome world of the infinitely small.

The properties of this world are closely tied to the properties of the infinitely large Universe. The brief information above can be regarded as something like the 'tip of the iceberg' that we see while considering processes at relatively low energies. The true nature of the phenomena in the microscopic world is much wider, of breathtaking interest and importance for cosmology. We will now have a close look at some aspects of the 'underwater' part of the iceberg. I need to warn the reader here that much remains unclear to experts in the structure of the submerged layers, and the deeper we penetrate into the iceberg, the more hypothetical some of the data will become. Nevertheless, this information is so important that I believe I must outline it to the reader, being of the opinion that the basic features of the phenomena have been determined by physicists rather correctly.

11

Grand Unification

When we were discussing the vacuum – the emptiness – in the chapter on 'Energy extracted from black holes', we emphasized that virtual particles are constantly created and annihilated in it. The emptiness proved to be a complex entity. The vacuum is a very complicated state of 'boiling' virtual particles of most different species.

The reader may not be too surprised by the statement that the properties of this state – the vacuum – depend on the recipe of its preparation. This implies that different vacua are possible; different types of emptiness!

In what follows, we will see examples of possible vacua. Now we will try to answer the following question: can the activity of the vacuum (its 'boiling') result in the formation of some energy density owing to the interaction of virtual particles?

Energy density can indeed appear. Zeldovich emphasized this fact in the 1960s. Each energy corresponds to a certain mass. Therefore, mass density will arise together with vacuum energy density. The reader may ask here: does it mean that some sort of universal medium, a new 'ether' is emerging in our notions? If this is true, such a medium can restore the concept of absolute rest and absolute motion. Indeed, the motion relative to this medium would be the motion with respect to the emptiness, in other words, with respect to the absolute space.

It may seem that if we move relative to this new 'ether', we should feel the flow going against us, the 'ether wind' blowing in our face. This is what Michelson attempted to detect even in the last century, trying to measure the motion of the Earth through the ether in the experiments that were described earlier and that, I remind, gave negative results.

If the new 'ether' resembled an ordinary medium, the 'face wind' would indeed be detectable. The thing is, however, that the vacuum is an extremely unusual medium. In addition to the energy density, stress appears in it, like stresses arising in a solid object in response to tension. This stress is equivalent to negative pressure; one simply says that negative pressure is produced.

In ordinary media, pressure and stress account for only a small fraction of the total energy density (which includes the rest mass). The negative pressure in the vacuum is huge and equals energy density in magnitude. This unusual property signifies the important dissimilarity of the vacuum relative to ordinary media.

When an observer starts moving in this medium, the oncoming energy flow does meet him and it may seem that the observer could measure this flow (this would be the 'wind'). However, another oncoming energy flow, due to negative pressure, will also be there. This flow has negative sign, its magnitude equal to that of the former flow and exactly canceling it. As a result, no 'wind' is produced. Whatever the motion of an observer by inertia, he always measures the same energy density of the vacuum (if it is nonzero) and the same negative pressure, so no 'wind' will be created by the motion. The vacuum is the same for all observers moving by inertia with respect to one another.

We shall often return to the vacuum but for the time being we leave it and turn to elementary particles.

I have mentioned already that the electromagnetic interaction between particles with electric charge is the result of photon exchange.

The weak interaction is also due to specific charges. The essential difference between the electromagnetic and the weak interactions is that the latter occurs only at very short distances. We have seen that this is caused by the enormous masses of the intermediate (carrier) bosons. The interacting particles may 'borrow' energy for the creation and transfer of carrier bosons for only a very short time. Therefore, they can interact in this way only by being very close to one another. What would happen if the masses of all intermediate particles, both photons and bosons, were zero? Or another question: what would happen at very high temperatures when bosons are created as easily as photons?

Indeed, all particles possess high energies at high temperatures, so there is no need to 'borrow' energy for creating massive bosons. They already possess this energy. The exchange of these bosons would then be as efficient as that of photons, so complete symmetry would manifest itself for the weak and electromagnetic interactions. It is found that under these conditions (i.e. at very high energies) these two interactions demonstrate their inherent unity and thus merge into a unified electroweak interaction.

At sufficiently high temperatures, therefore, the interaction between particles is the unified electroweak interaction (calculations show that this should happen at a temperature of a million billion kelvin). Its carriers – the already mentioned bosons and photons – are in abundance and have zero mass. It is found, however, that the mass is zero not only for the interaction carriers but also for all the particles mentioned before: quarks, electrons etc! In this sense, they become similar to photons. What happens then if temperature is decreased?

The obvious symmetry of the electromagnetic and weak interactions is violated and disappears. Why does that happen?

The thing is that new fields and their quanta come into play: the particles we know nothing about yet. These are the so-called Higgs particles, bearing the name of their inventor. These particles disrupt the symmetry. Were it not for the Higgses, all particles would remain massless and the symmetry of the electromagnetic and the weak interactions would survive. Before speaking about the Higgs fields and symmetry breaking for the electromagnetic and the weak interactions, however, I wish to refresh the reader's memory about one simple experiment.

Imagine a ball that can roll in a depression of spherical shape. Wherever we put the ball in that depression, it rolls down and, after some oscillations around the lowest point, stops at the bottom. The reader probably remembers that the higher we lift a load above the lowest possible position, the higher its potential energy is in the gravitational field (proportional to the height). Therefore, when a ball is somewhere on the slope of the depression, its potential energy is higher, the farther the ball is from the symmetry axis. The energy of the ball is lowest at the bottom; one says sometimes that the ball is at the bottom of the potential well.

So far everything is quite simple. Let us ask now if, in a symmetric depression, the ball will always settle at the symmetry axis? No, this is not so. Imagine a small hill at the very center of our depression. Wherever we place the ball now, it stops not at the central symmetry point but at the lowest points to the side of the central hill. Its position at rest will be quite non-symmetric, despite the perfect symmetry of the pit with a hillock.

True, if we place the ball precisely on the top of the central hill, it will stay in this symmetric position. This cannot last long, though, since this equilibrium position is unstable and the ball will roll off to the unstable non-symmetric position in response to even the smallest perturbation.

This example shows how an obviously non-symmetric equilibrium state arises in a completely symmetric system with a symmetric initial position (at the top of the hill). The moment of breaking of the symmetric state and the point where the ball stops are accidental and are said to occur spontaneously.

Let us return to particles and fields. Potential energy can also arise in their interactions. The amount of potential energy can by analogy be described by the position of the ball in a depression – the potential well. Depending on the situation, the well may or may not have a central tip. Of course, it may be difficult for the reader to make the connection between a field and a ball in a depression.

However, abstract images are widely used in science. In this particular case, the height of the ball above the bottom of the depression describes the potential energy of the field.

Let us return to the Higgs fields. They can exist in two states. At a temperature above a million billion kelvin, the fields exist as individual elementary particles. As temperature drops to this limit, the Higgs fields undergo what physicists call a phase transition; they 'condense' like water from cooled overheated vapor. This produces a 'condensate' of Higgs fields which is independent of either spatial position or time. It cannot be removed in any way under these new conditions. In other words, therefore, this is a vacuum. This is precisely what physicists say: the 'new vacuum' is created.

The position of the ball at the top of the central hill corresponds to the 'old vacuum'. At higher temperatures the shape of the well would be different: its slopes would rise immediately from the center, so this point would be a stable equilibrium position of the ball. 'Old vacuum' is sometimes called the 'false vacuum', or the 'vacuum-like state'. (We shall often use the latter term.) As temperature decreases, the well shape changes to one with a hill at the central point.

The formation of the new vacuum is equivalent to the ball rolling off to the lowest state, that is, to the lowest-energy position in the valley by the central hill. Its position is definitely not symmetrical. The state thus created is 'askew'.

The Higgs fields thus split into dissimilar components. One corresponds to a quantum, a massive zero-spin particle which is absorbed by the carrier particles, so that the vector bosons thereby become massive themselves (I will not try to explain why this happens). At the same time, particles of matter also gain masses: quarks, electrons etc. This happens because they interact with the non-symmetric condensate of Higgs fields which formed the new (non-symmetric) vacuum. Again, I choose not to explain why and how this occurs. This would be a very complicated explanation, and the reader may already be overloaded with outlandish information. Note that the photon, as the carrier of only electromagnetic interactions, remains massless.

These multifaceted consequences were brought about by the Higgs fields 'rolling off' at lower temperatures to the non-symmetric state of the new vacuum.

The details of symmetric and non-symmetric positions of the ball, of fields 'rolling off' energy hills may sound too abstract and far-fetched. Alas, this cannot be helped; the reader will have to apply certain attention and fantasy even in a very simplified explanation.

Now that the 'rolling off' has occurred, the carriers of the weak interaction gain masses. This makes the weak interaction act over a very short range, while the massless photon still ensures electromagnetic long-range interaction. The former symmetry is unrecognizable now. The symmetry that was so obvious at high temperatures is now broken and hidden.

This explains why it was so difficult for physicists to recognize this under the conditions of today's Universe. Still, they did it! Stephen Weinberg, Sheldon Glashow and Abdus Salam were given the Nobel prize for physics in 1979 for the creation of the unified theory of the electroweak interaction.

The theory of these processes at the very beginning of the expansion of the Universe, at enormously high temperatures, was proposed by the Moscow theorist David Kirzhnitz. Later he continued to work on the theory together with Andrei Linde, his young colleague.

Not all details of the picture outlined above have been confirmed to the same degree of reliability. For example, the search for the massive Higgs particles has not been successful so far. At least one species of this heavy particle must survive after the evolution suggested by the theory, and it must exist in today's Universe. Experimental detection of this particle will be extremely difficult but physicists believe in the ultimate success of the search.

Let us turn now to the strong interactions. Particles undergoing the strong interaction – the quarks – and those that do not feel it – for example, electrons – appear completely different in this respect and their mutual transformations seem to be impossible.

I have mentioned already that the unified electroweak interaction sets in at temperatures above a million billion kelvin. At a lower temperature, it splits into the electromagnetic and the weak interaction. Superficially, these are very dissimilar interactions. The strong (color) interaction keeps very much away from these two even at these high temperatures, and does not resemble the electroweak interaction at all. Whereas all particles take part in electroweak interactions, only quarks undergo the strong interaction.

None of the processes that we have treated so far can result in the transformation of, say, a quark into an electron or a quark into an antiquark. Of course, collisions of sufficiently energetic electrons may create quarks as well but only in pairs with antiquarks, so that the total numbers of the two counterparts are identical. Likewise, a collision of a quark with an antiquark results in their annihilation and transformation into other particles, but it is always a pair that disappears, never a single quark or an antiquark.

What is conserved in nature, therefore, is the difference between the number of quarks and that of antiquarks. This difference is known as the baryonic charge (to be precise, the difference divided by 3). Until now, the baryonic charge was conserved in all physical experiments. Is it possible, however, that at very high energies, much higher than those we have already considered (and they were very high indeed), reactions are possible in which the baryonic charge is not conserved and which are impossible at lower energies (so that physicists could not discover them)?

The theory concludes that such processes are possible but only at fantastically high energies.

We were discussing energies of particles at a temperature of a million billion kelvin. Now we need to look at temperatures and energies still greater by a factor of a thousand billion. What should happen at energies so much higher?

Note, first of all, that the higher the energy, the smaller the distance to which the colliding particles can approach one another.

It is shown that at the minute distances that are a million billion times less than the size of the atomic nucleus (hence at energies corresponding to a temperature of a billion billion billion kelvin) all three types of interactions – electromagnetic, weak and strong – must become equally efficient and shed their individuality. At energies above this value there must exist the Grand (universal) interaction.

At such high energies, new particles are born in copious amounts: particles that carry the universal interaction. Their masses are a thousand billion times higher than those of the intermediate bosons discussed in connection with the electroweak interaction. Particles that heavy can only be created at still higher energies. We never mentioned these particles before because then we stayed in the realm of substantially lower energies.

The properties of the carriers of the universal interaction are truly amazing: they can transform quarks into other particles and vice versa, and quarks into antiquarks. Now the differences between quarks and such particles as electrons or neutrinos, which were so pronounced at low temperatures, fade away and they all appear as distinct manifestations of the same 'superparticle'. This disappearance of differences signifies the arrival of a new, higher order of symmetry: the symmetry of Grand Unification.

In addition to the particles we have met before, another set of Higgs particles exists at the very high temperatures discussed above (these Higgs particles are not identical to those we were concerned with earlier). As temperature decreases below the Grand Unification threshold, the already familiar Higgs mechanism is triggered which breaks the symmetry, this time the symmetry of the Grand Unification. The difference is that this time it happens to the new Higgs particles.

At temperatures above the Grand Unification point, Higgs particles are free. As temperature decreases, a new condensate of the Higgs field is formed: the new lowest-energy state of the system, that is, another species of the vacuum, the third one in our book.

Different vacua, or rather different 'vacuum-like states', possess different energy densities. As a result of formation of the Higgs condensate, the carriers of the universal interaction gain masses: they become superheavy, and cannot be created at low temperatures. The unified interaction now splits into the strong and the electroweak interactions.

We have thus seen that as energy (and temperature) increase, different types of interaction, all very dissimilar under ordinary conditions, acquire similar features and then merge into a unified interaction.

Einstein's dream is thus coming true in our lifetime: the dream of unifying all forces in the Universe. Three forces fuze into one at the energies of Grand Unification: the electromagnetic, the weak and the strong. The only force that has kept apart so far is the gravitational force which acts on all species of matter. Very little is left undone: to unify – at some absolutely superhigh energies – the gravitational force with the already unified force of the Grand Unification. Alas, this last step has proved to be the most difficult for the theory.

Before turning to the latest attempts of theorists to unify the gravitational force with the other forces in the Universe, one has to recall that the nature of the gravitational field is essentially geometrical: this is the curvature of spacetime. It is necessary to add to this that under appropriate conditions the gravitational field can, as the electromagnetic field, manifest quantum properties.

The reader remembers that the quanta of the electromagnetic field are photons. The quanta of the gravitational field are gravitons – the as yet undiscovered hypothetical particles – which act as carriers of the gravitational interaction. Like photons, gravitons have zero rest mass and always move at the speed of light.

Albert Einstein strongly believed that the electromagnetic field must also have a geometric nature. He devoted the entire second half of his life to attempts at finding a geometric representation of the electromagnetic field which, as he hoped, determines the macroscopic properties of matter. On one side of his gravitation equations we find quantities describing the curvature of space-time (the so-called curvature tensor) and on the other side we find the source of gravitation and curvature, that is, quantities that describe matter and non-gravitational fields (the so-called energy–momentum tensor for matter).

Einstein believed that this duality must be alien and unnatural for the final theory. If the left-hand side consists of geometric quantities, then the right-hand side must have quantities of essentially the same physical nature: geometric. For Einstein this meant that the description of matter and fields must be in terms of geometry. Infeld recalled how Einstein described the disparity: '... relativity rests on two columns. One of them is powerful and beautiful, as if made of marble. This is the curvature tensor. The other is rickety, as if made of straw. This is the energy–momentum tensor...We will have to leave this problem for the future.'

After more than three decades of work, Albert Einstein thought that he was near its final solution. In 1945 he wrote to Infeld that he hoped that he had discovered how gravitation and electricity were related to each other, although he felt that the physical justification was still a long way off. In his attempts to unite gravitation and electricity, he additionally introduced 'twisted' spacetime with which to describe electromagnetic phenomena. Alas, these specific efforts proved to be unsuccessful and no unified theory was developed.

In the 1920s the German physicist Theodor Kaluza and the Swedish physicist Oskar Klein made a new attempt to unify Einstein's gravitation and Maxwell's electromagnetism on a geometric basis but with a very different approach. They assumed that spacetime is far from four-dimensional (three spatial coordinates and one temporal coordinate) but has a fifth spatial dimension, which they introduced. They wrote equations for the curvature of the five-dimensional world, similar to Einstein's equations for the four-dimensional world. It was found that the additional equations that arise because of the presence of an additional dimension are the equations of Maxwell's electrodynamics. It was thus found that electromagnetism can also be geometrically interpreted, although the interpretation is very unusual: it is connected with the fifth dimension.

Kaluza and Klein's attempt could not, in fact, be regarded as an unqualified success. In addition to a number of difficulties that we will not discuss here, their theory poses an obvious problem: why is it that the additional dimension does not manifest itself in any real way in our world? Why are we allowed to move in space in three directions (in length, width and height) but not in this additional, hypothetical dimension?

To remove this difficulty, Kaluza and Klein had to introduce a number of very artificial assumptions which were essentially meant to forbid motion in the new dimension.

To recapitulate, the first attempts at unifying the forces of nature can only be regarded as mostly preliminary. We know that by mid-century many physicists treated them with extreme skepticism.

Let us return, however, to our time. I have described earlier how physicists achieved the understanding of the unified nature of different forces at high energies. This involved geometric ideas too: the ideas of symmetry. However, this was symmetry not in real physical spacetime but in an imagined abstract space that represents different states of particles and fields, that is, in the abstract space describing the internal characteristics of particles.

Now that we turn to the idea of unifying all forces with gravitation, we need to remember that gravitation is related to the curvature of real spacetime. Therefore, when constructing the superunification, we need to somehow unify the geometrical characteristics of the four-dimensional spacetime with the characteristics of the space of internal states. How could this be done? What is the purpose of this exercise?

Before beginning this story, I will point to another factor. When discussing particles, we have been dividing them into two large classes: particles of physical matter and particles mediating the interactions. The particles in these two classes have very different properties. When we were discussing interactions, these two sorts of particles served completely dissimilar functions. The intermediate particles mediated the interactions, as if they were 'servicing' the particles of matter. There was no chance for the two sorts of particles to turn into one another.

If, however, we think of a universal unification of all kinds of interaction into some unified interaction, the thought comes to mind of whether matter particles and interaction carriers can also be unified into some common entity. If it were possible, both matter particles and interaction carriers would be its different manifestations. Now that we know that modern physics unifies such dissimilar things as space and time, or electromagnetism and nuclear forces, the idea of unifying the constituent parts of matter and force carriers does not appear to be too absurd.

Furthermore, it was found that the unification of the gravitational forces with all other forces incorporates the unification of matter particles and interaction carriers and the possibility of their mutual transformations.

Of course, this supersymmetry of all forces and all particles can manifest itself only at very high energies, while under ordinary conditions it must be thoroughly hidden and violated, so that matter particles and interaction carriers, as well as different forces, do not resemble their counterparts. How high could be the energies at which the unifying nature of all fundamental interactions becomes apparent? They are found to be a hundred thousand times higher than the energy of Grand Unification. This energy is known as the 'superunification energy'. It corresponds to a temperature of a hundred thousand billion billion degrees.

I will only give very brief remarks on some recent versions of superunification. There are several reasons for this. First, it is extremely difficult to offer explanations without formulae, and this insert will have to be very short, since the aim of this book is somewhat different. Second, experts are not at all sure yet that they have grasped even the principal features of the phenomena, and the research continues in various directions.

The reader remembers the attempt by Kaluza and Klein to unify gravitation and electromagnetism into one object: they needed to introduce an additional dimension.

The task now is to unify gravitation with all types of forces and particles. An idea suggested itself to achieve this by introducing new additional spatial dimensions. This idea proved to be extremely fruitful. At present we know versions of the theory with 10, 11 and even 26 dimensions instead of the ordinary four dimensions of spacetime. (The theory that seems to be the favorite among them is the one with 10 dimensions.)

The geometric properties of these additional dimensions allow us to describe all manifestations of properties of matter interaction carriers in terms of a unified set of concepts. This realizes the grand dream of Albert Einstein.

The question that we asked earlier still remains: why is it that in real situations in our world we fail to detect these dimensions, that is, why can't we move in these dimension, as is often described in sci-fi novels?

The way to overcome this difficulty is outlined by the idea of *compactification*. According to this idea, the additional spatial dimensions are twisted and closed (as one of the dimensions of a sheet rolled into a cylinder). These additional dimensions are compactified when energy drops below the Planck energy. Note that the radius of the contracted dimensions is absolutely infinitesimal: it equals the Planck length mentioned above. This length is a hundred billion billion times less than the size of the atomic nucleus.

Obviously, the negligible extension of additional dimensions does not allow the detection of these dimensions under ordinary conditions of relatively low energies. They make themselves felt only via the variety of forces and charges of particles.

Supersymmetry assumes the existence of a large family of new particles. None of these particles have been discovered yet.

Theories with very complicated and exotic sets of particles are known now. Alas, I have to cut short our exciting journey into this area which is still largely unexplored.

Our brief visit to the wonderful microscopic world will now allow us to throw a glance at what happened at the very beginning of the expansion of the world, that is, explore how our Universe exploded.

The two preceding chapters discussed at great length the achievements of modern physics and astrophysics but turned very sparingly to the concept of time as such. At first glance, but only at first glance, this seems rather strange in a book whose protagonist is time. The striking properties of time, which disclose its profound

187

meaning, are discovered in the processes which occur in the depths of the microscopic world and in the vast expanse of the cosmos. We can continue with our story of time only after an adequately detailed acquaintance with these processes.

12

Sources

We are now departing on a voyage to the very sources of the river of time. What was it that happened at the very beginning of time? What triggered the expansion of the Universe?

We have seen in the chapter 'Towards the sources of the river of time' that the huge pressure of hot matter at time zero cannot be the cause of the high velocities of recession of matter, because the uniform Universe has no pressure drop, which is the only cause of force driving an expansion. What then was the cause of the expansion?

The key to understanding the 'primeval push' lies in the existence of the special vacuum-like state of matter at high densities and temperatures.

We have already looked at several vacuum-like states in the chapter dealing with Grand Unification. Theorists believe that a unique vacuum-like state with enormous energy density and the corresponding gigantic mass density is formed at the temperature of 'superunification'. This density in grams per cubic centimeter is written as unity with ninety four zeros (!). The enormity of this number defies imagination. We have already mentioned in the preceding chapter that any vacuum possessing non-zero mass density must have huge negative pressure.

In accordance with Einstein's theory of gravitation, gravitation is produced not only by mass but by pressure as well. Pressure is usually not high and so the gravitation connected with it is negligibly small. In the case of the vacuum-like state the picture is very different since pressure is huge and the gravitation it creates is greater than that produced by mass. However, the pressure of the vacuum is negative, so that instead of gravitation it produces antigravitation – gravitational repulsion! This is the gist of the matter. This phenomenon is the key to understanding the 'primeval push'. Given the gigantic initial density and temperature (superunification density and temperature), antigravitational forces result in overpowering repulsion of all particles of matter. These particles gain enormous initial recession velocities. In view of this, the process of superfast expansion of the Universe was called 'inflation'.

It is also very important that the primeval vacuum-like state was extremely unstable. It existed only for about one hundred million billion billion billionth of a second! Then it decayed, and its mass density transformed into 'ordinary' super-elementary particles with enormous energies (those we discussed in the previous chapters). This is how the vacuum-like state gave birth to the hot Universe at a temperature, at that moment, of a billion billion billion kelvin.

Particles born of the primeval vacuum had high initial velocities of recession, owing to the forces of antigravitation. However, as the 'supervacuum' decayed, these forces disappeared and were replaced by ordinary gravitation. The newborn hot matter, flying outward, became very rarefied and cold many billions of years later; it fragmented into pieces from which galaxies, stars and systems of stars were formed at later stages. The physical processes taking place during this evolution are described in detail in a number of books, including some popularizing books. I will, therefore, be very brief about these topics.

Once the 'false vacuum' had decayed and the Universe heated up, there was a very special superhot plasma of elementary particles and their antiparticles of all possible sorts. They all interacted with one another very intensely.

As the Universe expanded, it was cooling down. About one tenth of a second after the onset of expansion, the temperature had dropped to thirty billion kelvin. The hot matter included a large number of high-energy photons. Their density and energy were so high that light interacted with light, generating electron–positron pairs.

The annihilation of these pairs produced photons and also

neutrino–antineutrino pairs. Ordinary matter was also present in this violent 'cauldron', but at very high temperatures complex atomic nuclei could not survive: they were immediately broken by the surrounding energetic particles; therefore, matter existed in the form of neutrons and protons. Constantly interacting with energetic particles of the cauldron, neutrons and protons rapidly transformed into each other but were unable to join into a nucleus since high-energy particles around them immediately smashed them apart. This is why the chain of events that would lead to the formation of helium and other heavier elements was cut short by high temperature at the very beginning.

Several minutes after the start of the expansion, the temperature of the Universe had dropped below a billion kelvin. Now neutrons and protons could join and form deuterium nuclei. The newborn deuterium nuclei entered a further chain of nuclear reactions, until helium nuclei were formed. This was the final stage of nuclear synthesis in the early Universe.

Calculations have demonstrated that the primeval matter had to contain about 25% helium, with the remaining 75% consisting of hydrogen nuclei (protons). Observations have confirmed that the oldest stars in the Universe had a composition supporting the predictions of the theory of the hot Universe. Nuclei of still heavier elements were synthesized in the Universe by nuclear processes inside stars, but much later (in the epoch nearer to ours).

Nuclear reactions in the early Universe stopped five minutes after the start of the expansion. By this time all active processes with elementary particles had been completed; nothing 'of interest' took place in the Universe for a very long period that followed.

All this time the expanding matter was kept ionized because of high temperature (such matter is known as *plasma*). Dense plasma is opaque to radiation, so that radiation determined the force of pressure. Density oscillations of small amplitude (acoustic waves) were propagating through this mixture of plasma and radiation. No other processes were occurring in the expanding matter.

Only after 300 000 years of this dull existence had the plasma cooled to 4000 degrees and turned into neutral gas (atomic nuclei had captured free electrons). This gas became transparent to the primordial radiation. Now its pressure was determined only by the motion of neutral atoms (the pressure of radiation disappeared), the elasticity of the gas dropped dramatically and the mechanism of the so-called gravitational instability became important. The theory of these processes was developed by the Moscow physicist E. Lifshitz in 1946.

The higher-density clumps of acoustic waves which had considerably large linear sizes were then more and more enhanced by gravitational forces. Finally, these higher-density areas formed large clouds which then evolved to galaxies and clusters of galaxies. Stars arose within galaxies.

This, however, is another story. Let us go back to the very beginning.

We looked at how the vacuum-like state generated the primeval push. This, modern astrophysics tells us, is the secret behind the mystery of the birth of the Universe.

The first 'hunch' that the vacuum-like state and hence, antigravitational forces, may arise in superdense matter at the start of expansion of the Universe was formulated by E. Gliner, a physicist in St Petersburg (then Leningrad). He came to Moscow at the end of the 1960s to outline his hypothesis to the 'high and mighty' of cosmology and related sciences. Alas, he was not understood. As all the others, I did not understand anything either. I was of the opinion that giant negative pressure cannot materialize in nature, so thinking about antigravitational forces was meaningless. Almost everybody reasoned that way, and I lacked fantasy too. However, two Moscow physicists, David Kirzhnits and Andrei Linde, showed at the beginning of 1972 that a state of this type can indeed arise in the expanding Universe. Somewhat later these ideas were developed and applied to cosmology by E. Gliner, L. Gurevich and I. Dymnikova in Leningrad and then, using the latest achievements of high-energy physics, by A. Guth, A. Albrecht and P. Steinhart in the USA and by A. Linde, A. Starobinsky and others in Russia.

A number of questions arise at this point, the first of them being: 'What was there before all this?'

This is a difficult question. There was no answer to it even as recently as several decades ago. Furthermore, some of the Soviet philosophers at the time when I started doing science regarded the question as anti-scientific and anti-Marxist. 'Nonsense! Do you say that the Universe had a beginning? Therefore, that it was created by God?' and so on, and so forth. I happened to be in a very unpleasant mess soon after I started working in the group of Academician Yakov Zeldovich.

A journalist from the Moscow daily *Komsomolskaya Pravda* (published by the Young Communist League) visited our laboratory and requested an interview about the problems of modern cosmology. We had a talk for some time and the journalist left, very much excited with the problem of the onset of the expansion of the Universe (almost nothing was reliably known about this at the time); he promised to prepare a draft article for his paper. We all know that the style of 'about science' writing in newspapers is frequently a far cry from the style characteristic of scientists. The paper did not bother to check the final text of the interview with us and the article was published under the 'flashy' heading 'When the Universe did not exist yet'. Of course, a number of important aspects were misunderstood and misinterpreted, while the unexplained assertiveness of the heading led to trouble.

We were attacked by 'materialistic' philosophers but also by the functionaries who regarded this onslaught as their 'duty'. Consequences of the pressure of the ideological watchdogs of the 'purity' of dialectic materialism could have been very serious were it not for the unique position of Professor Zeldovich – Academician and three times 'Hero of Socialist Labor'. Of course, the progress of science cannot be stopped by bans or intimidation. After a great deal of research over the last quarter century, the questions about the very beginning of the Universe and the 'What was there before?' problem are gradually being answered and clarified.

As I have said already, the expansion probably started with a superdense vacuum-like state and tremendously high temperature. The curvature of spacetime and the tidal forces it describes are as high as in the singularity of a black hole. The singularity at the beginning of the Universe (known as the cosmological singularity) is similar in many ways to the singularity in a black hole. However, there are also important differences. First, the cosmological singularity occurred in the entire Universe, not in some part of it, as in the case of a black hole. Second, we find it not at the end of the contraction phase (as the black hole singularity) but at the start of the expansion process.

This last factor is especially significant. We cannot see the singularity of a black hole from the outside of this black hole, and it does not affect in any way the events in the Universe outside the black hole. (The British theorist Roger Penrose called this feature the 'cosmic censorship principle'.) In contrast to this, the cosmological singularity was the source of all processes in the expanding Universe. Everything we observe today is a consequence of singularity. In this sense we can study the cosmological singularity by observing its consequences: we can 'visualize' it.

The Moscow physicists Vladimir Belinsky, Evgeny Lifshitz and Isaac Khalatnikov found the most general solutions of the equations that describe the possible motion of matter close to the singularity. Everything we have said about the singularity in black holes is true of the cosmological singularity. What was there before the singularity? Was the entire matter previously compressed and was the ordinary time ticking?

We still do not know the ultimate answers to these questions. Most specialists are of the opinion, nevertheless, that there was no compression phase and that the cosmological singularity was the source of the river of time in the sense in which the black hole singularity is the sink for 'time rivulets'. This means that time in the cosmological singularity also decayed into quanta, so that the question 'What was there before the singularity?' becomes meaningless.

Much in this field remains uncertain. It is probable that something like a 'foam' of spacetime quanta existed close to the singularity, on the space and time scales outlined above; physicists say that space and time undergo quantum fluctuations. Tiny 'virtual' close worlds and virtual black and white holes are born and immediately disappear. This microscopic 'boiling' of spacetime is in some respect similar to the creation and annihilation of virtual particles that we discussed when describing the quantum nature of the vacuum (see p. 140).

The reader will also recall that at such high energies on a very small spatial scale, the space may have more than three dimensions (see pp. 186–187). These additional dimensions remain 'rolled in', 'compactified', while in the three spatial dimensions the Universe expands and transforms into what we know as 'our Universe'.

This is the range of problems that strongly attracted Andrei Dmitrievich Sakharov in the first half of the 1980s. Thus he discussed the possibility for the Universe to be created by quantum processes from exotic states of matter in which time has not one dimension (as in today's Universe) but two, three etc. (i.e. time had 'length', 'width', 'height', ...), and even from states that only have space (of more than three dimensions) but have no time.

Sakharov also hypothesized that in the extremely small regions of our current Universe that can be accessed for investigation only by particles of tremendously high energy (far beyond the limit of the latest accelerators), time has many dimensions and exists 'twisted' into supercompactified braids; these 'braids' manifest themselves in very specific properties of elementary particles. Was our time created together with the Big Bang or did time already exist when the Universe was yet unborn? I had a curious conversation concerning these questions in the autumn of 1988 with two well-known scientists: R. Ruffini whom we have already met and G. V. Coyne, Director of the Vatican observatory.

The reader should not be surprised that there are astronomers in the Vatican, engaged in the most advanced research in astrophysics, and that we meet them and discuss numerous problems.†

One of the best known cosmologists, the Belgian Georges Lemaître, who greatly contributed to the development of Friedmann's theory and established its connection to observations, held the position of President of the Papal Academy in the Vatican in 1960–1966.

Today's world is very complex and multifaceted, and is becoming gradually more open and interrelated. Pope John-Paul II is an active supporter of peace on the entire globe. Some time ago he had a reception for representatives of the astronomical and space exploration communities. I felt slightly uncomfortable, listening to the Pope's speech in the reception hall in the Vatican. The Pope called for peaceful exploration of outer space, unified effort of all peoples and further successful progress of science. The speech caused a complicated knot of feelings, emotions, knowledge and faith. As far as I know, my colleagues had a similar response.

When G. V. Coyne and myself met in Moscow, I decided to ask

† I wrote this sentence for the benefit of a Soviet reader who had been persuaded for dozens of years that religion cannot have anything in common with true science.

him what he thought of the concept of time, why is it flowing from past to future. I was not surprised that my question somewhat baffled him: almost everybody is baffled when confronted with this question. Indeed, this is one of the simplest 'childish' questions that are especially hard to answer (see 'Preface to the Russian edition'). After a short pause Coyne remarked that although he could repeat the well-worn statements from physics that I was certainly familiar with (we will have a look at them a bit later), he preferred to attract my attention to the thoughts that Saint Augustine had about time. I have already quoted Saint Augustine in the preface. Coyne reminded me that Augustine always insisted that time was created together with the Universe. Therefore, the question of what was there before the Universe was born is meaningless, since there was no 'before' and no time either. This was a very profound remark.

From the standpoint of today's knowledge, we should say that time drastically changed its properties in the singularity, and the moment of the start of the expansion was the source of our continuous time flow. Can we say anything else about the superdense singular state? The American physicist John Wheeler has been working for several decades on an idea that the space and time in this exotic state decay into quanta. John Wheeler is one of the patriarchs of modern theoretical physics. But even in advanced age he is wonderfully active, travels all over the globe, takes part in conferences, talks to colleagues and in addition, he has never lost his wonderful spontaneous humor. In summer 1992 he paid a visit of a couple of days to Copenhagen where I now live. The visit was not unexpected, of course. I had been almost knocked down when I read his letter seven months before this meeting, asking me whether I might be free for approximately one hour at 9:30 on Friday the 5th of June 1992, since he would much like to see me. That the time was so exact was no joke, Wheeler was dead serious. I wrote in my reply that I would try to be available, even though I had no idea of what was going to happen in half a year's time, and that I could only envy and applaud a man who was still so busy and capable of scheduling his time more than half a year ahead.

We did meet, and had a lively discussion of the current problems of black hole physics, since at that time my friend and colleague Valeri Frolov and myself were working on a new edition of our monograph devoted to this topic. Before Wheeler left, I asked him: 'John, you pioneered several revolutionary developments in physics and in addition you are famous for your pithy, terse definitions of the most profound concepts of modern physics. Could you try and formulate what time is? I need it for a physics-popularizing book, to be translated into English.'

John took a very long time to mull it over; I suspected that he had fallen asleep (we had just finished a very good lunch). Actually, he was deep in thought. Opening his eyes, he said very seriously: 'I will think about it and write to you'. During the month that followed, John visited several more places in Europe; having returned to the USA, he did not forget about my request; here is what I got in his letter (sent together with a copy of his book *Frontiers of Time*, with a handwritten dedication: 'To Igor – may you be timeless! – John. 25.IX.92'): 'You asked for a phrase. There are graffiti on the wall of the men's room in Austin, Texas, and among them is this, "Time is nature's way to keep everything from happening all at once".'

Let us return to the 'quantum foam' of space and time on the singular state of the Universe. This is a complex conglomeration of arising and immediately disappearing black and white holes, very small closed mini-Universes and more complex structures and their complexes. One of the hypotheses developed by the physicist we already know, Andrei Linde (who worked in Moscow but is now Professor of the Stanford University, USA), is a breathtaking scenario of one of the tiny bubbles of the 'quantum foam' growing into our Universe. An accidental fluctuation in this bubble causes a random drop of the density to a value much lower than the initial level. At this reduced density (even though still unimaginably higher than our ordinary values) the gravitational repulsion of the vacuum dominates the random quantum fluctuations, so that bubble grows catastrophically and turns into 'our' Universe. One has to remember, though, this is a very rare event and the predominant part of the bubbles created die out immediately, returning to the initial state. The reader remembers that the time scale of these events is quite infinitesimal:

> Worlds on worlds are rolling ever
> From creation to decay,
> Like the bubbles on a river
> Sparkling, bursting, borne away.
>
> Percy Bysshe Shelley

According to this hypothesis, universes are very rarely created explosively in this 'breathing foam'. They probably differ from one another in all their properties, including the number of spatial dimensions, the properties of time, laws of nature ... We live in one such universe, where conditions for the evolution and existence of intelligent creatures were realized accidentally.

I need to emphasize that the main part of the 'pre-matter' beyond the boundaries of our bubble (our Universe) still remains, as it always has been, in its 'quantum boiling'. The reader should not forget the conditionality of the notions of 'now' and 'always' when speaking in terms of time quanta.

This is thus the world by A. Linde: the picture of eternal creation (and death after collapse) of new universes, the picture of the exploding Eternity.

The expanding Universe discovered by A. Friedmann and Edwin

Hubble, 'our Universe', which only recently seemed to be over-whelmingly complicated as something defying human imagination, looks now more like a tiny sand grain in the stormy flow of time, rushing along a tortuous and treacherous bed.

13

What produces the flow of time and why in a single direction only?

Contemporary science has uncovered the relation between time and physical processes, making it possible to 'grope' for the first links of the time chain in the past and to project its properties to the distant future.

But what does modern science say about why time flows at all, and why only from the past to the future? I should immediately say that experts still lack an exhaustive, clear and generally accepted answer to this question. Nevertheless, a great deal has been achieved in this field, too, and we will have a quick look at some fragments of the achievement of the science of time.

In the post-Newtonian era, physicists have always emphasized a surprising property of the laws of nature: they do not in any way single out the direction of time flow from the past to the future.

We easily recognize this fact by looking at the simplest problems in mechanics. For example, let a ball roll along a surface, hit a wall at a certain angle, rebound and continue rolling. Now we can, in our minds, reverse the direction of time and imagine the ball rolling in the opposite direction, going through all the points of its trajectory in the opposite order. It is as if we had filmed the experiment and then projected the film beginning with the last frame. All laws of mechanics describe the motion of the ball equally well both in the forward and the reversed directions of time flow.

Another, this time a more complicated, example. Consider a planet rotating around the Sun according to the laws discovered by Johannes Kepler. If we reverse the flow of time (physicists say that we reverse the sign of time from 'plus' to 'minus'), we get a planet moving along the same orbit but in the opposite direction. Kepler's laws would be perfectly obeyed.

The laws of Newton's physics thus describe the forward and reverse motions equally well, making no distinction between them.

These laws do not dictate the flow of time from the past to the future. Physicists call this property the T-symmetry or T-invariance. The same property is shown not only by Newton's laws but also by the laws of electromagnetism, and of special and general relativity.

T-invariance allows one to calculate events both towards the future and towards the past. For example, the laws of celestial mechanics serve to calculate motion in the future, and hence future sightings of Halley's comet in our skies; with the same precision, we can calculate the dates when this comet approached the Sun and the Earth in the distant past. Observations confirm the correctness and accuracy of these calculations.

In the 18th century and the first half of the 19th century people were mostly certain that all processes in nature can ultimately be reduced to mechanical motion and the interaction of particles, their attraction and repulsion. In this case, the laws controlling these motions can in principle be computed arbitrarily far both forward and backward in time.

The past and the future of the Universe could then be calculated with equal certainty. They are completely predetermined by the positions and velocities of all particles in the Universe fixed at some moment of time – this was the viewpoint of the founder of celestial mechanics, the French astronomer and mathematician Pierre Simon de Laplace (1749–1827). He wrote that one

... must consider the present state of the Universe as a
consequence of its previous state and as a cause of the
subsequent one. The intelligence which at some moment
of time would know all the forces acting in nature and the
relative positions of its parts, and which would be powerful
enough to subject this knowledge to analysis, would
encompass in one formula the motion of the greatest bodies

in the Universe and the motion of the lightest atoms; there
would be nothing left unknown and both the future and the
past would be open to its gaze.

P. S. Laplace, (1820) *Theorié Analytique de Probabilité*, (Paris: V.
Courrier) (Tranlated from the Russian version by V. Kisin)

The St Petersburg physicist A. Chernin remarked: 'Our contemporary who chose to support Laplace's view, could say that the future is as if recorded on a roll of cinema film which – in its final form – is being unrolled and projected to us. The film can be viewed both forwards (to the future) and backwards (to the past).'

What is this then: the laws of nature do not distinguish between the past and the future? Do they allow motion with equal freedom in either direction of time, the past or the future? Why is it then that time moves in one direction only? We know exactly that this is so for time. We remember the events of the past. Even remote events in the past leave traces in our memory. However, we remember nothing about the future! The past is behind us, it cannot be changed in any way, while the future can be influenced. We know all this both from the accumulated knowledge of science and from our everyday experience. There is no symmetry in the flow of time in nature; physicists say that time is completely anisotropic. However, this is in no way reflected in the laws of motion of matter.

In fact, I have to add one very important qualification.

In 1964 the American physicists J. Cronin and W. Fitch discovered a process that was not T-invariant. In other words, this process is sensitive to the direction of time flow. Sixteen years later Cronin and Fitch received a Nobel prize for this discovery.

They found that the decay of an unstable particle, the neutral K meson, occurs in such a way that it feels the direction of time flow. Could that be the explanation? If there exist processes that are not T-invariant, could they determine the direction of time flow and the rate of flow as well?

Unfortunately, the discovery made by Cronin and Fitch is unlikely to give a solution to the mystery of time. Time-asymmetric processes of a very specific class occur very rarely and only to exotic particles. However, we know that the directed flow of time is evident in everything and always, whatever happens in the Universe. It is impossible (at least today) to draw a picture of how rare and exotic processes can control this ubiquitous directed time flow. OK, rare processes are not the factor. But what is?

Stressing the reversibility of time in elementary processes, physicists established quite long ago that time is irreversible in complex processes; such processes were accordingly called irreversible. This was understood as early as the last century. Let us consider a simple case of a droplet of ink added to water in a jar. The droplet spreads out rapidly, so that the color becomes uniform in the entire vessel. Anyone can observe this phenomenon. However, no one has ever seen a process developing in the opposite direction: ink particles collecting from the whole volume into a single droplet. Why is this so? The laws dictating the motion and interactions of water and ink molecules are T-invariant, aren't they? If all particles of ink and water in the vessel are at some point given velocities that are exactly opposite to those they had, and all external influences are canceled, then all the events in the vessel are replayed in time in reversed sequence, so that the ink collects itself into a single droplet. Hence this picture is possible!

Yes, it is possible in principle but never happens. The thing is that ink gathering to form a droplet is, though possible, an incredibly low-probability event. Before going into details, let us consider an experiment demonstrated to high school students at physics classes.

Take an iron rod, heat it and then put it into a vessel with cold water. The rod will cool down, the water will get warmer and their temperatures will become equal. The process always goes this way.

Heat is never transferred from cold water to hot iron, raising its temperature still further.

But why is this impossible? Heat transfer from a cold to a hot body does not violate the law of conservation of energy. The thermal energy is conserved even though transferred from one body to another. However, the transfer goes, for some reason, in one direction only: from a hot to a cold body.

This is another example of an irreversible process, similar to the spreading of a droplet. These examples are very similar. Indeed, we know that heat is the chaotic motion of molecules. Therefore, if the velocities of all molecules in the water and the iron rod are switched to the oppositely directed velocities, and we again exclude external factors, then the process will develop in the time-reversed order (indeed, the motion of molecules is described by T-invariant laws!). Hence the heat will flow from a cold to a hot body. This, however, never happens in the real world.

Why does irreversibility always arise in all such processes, even though they are composed of particle motions that are definitely time-reversible? Where and how does reversibility perish?

This puzzle had already been solved in the 19th century.

In 1850 the German physicist Rudolf Clausius and in 1851, independently, the British physicist William Thomson, Lord Kelvin, discovered the law known as the second law of thermodynamics. This law was essentially a generalization of the experiments described above and could be formulated as an inference, stating that heat always flows from a hot to a cold body (I will describe the observations on a droplet of ink a bit later). Here is how Thomson formulated this law: there can be no process in nature whose only result would be mechanical work done at the expense of cooling of a heat sink. This statement immediately implied that the complete conversion of heat into mechanical energy or other types of energy is impossible. This means that if a system is isolated, then ultimately all types of energy in this system will transform to heat, the heat will spread uniformly over the system and the so-called thermodynamic equilibrium will set in.

We know perfectly well how this law manifests itself in real situations. For example, friction in mechanical systems converts some mechanical energy into heat. In heat engines, however, we can convert heat into mechanical energy but only if we maintain a temperature difference between the heater and the heat sink of the machine, otherwise it will not work. This means expenditure of energy, which partly converts into heat. The amount of heat produced in this process is greater than that converted into mechanical energy by the heat engine. This leads to the never ending accumulation of heat to which all other energies convert. Clausius later gave a mathematical expression for the process.

The thermodynamic ideas of Clausius and Thomson were then developed and extended by Ludwig Boltzmann. He uncovered the meaning of the second law of thermodynamics. Heat is, in fact, the chaotic motion of atoms and molecules of which material bodies consist. Hence the transition of the energy of mechanical motion of individual constituents of the system into heat signifies the transition from the organized motion of large parts of the system to the chaotic motion of the smallest particles; this means that an increase in chaos is inevitable owing to the random motion of particles, unless the system is influenced from outside so as to maintain the level of order.

Boltzmann showed that the measure of chaos in a system is the quantity introduced by Clausius: entropy. The greater the chaos, the higher the entropy. The transition of different types of motion of matter into heat means that entropy grows. When all forms of energy have transformed into heat, and this heat has spread uniformly through the system, this state of maximum chaos ceases to change with time and corresponds to maximum entropy.

This is the gist of the matter! In complex systems consisting of many particles or other elements, disorder (chaos) inevitably increases as a result of the random nature of numerous interactions. Entropy is that very measure of the degree of chaos. Of course, chaos is intensified only if no special measures are taken to maintain the degree of order. But then the system must be monitored and the process must be influenced from the outside. Therefore, when discussing examples, I stressed the absence of such external factors.

In the case of the hot piece of iron and cold water, the probability is much higher for the molecules of the hot iron rod, having higher energy, to transfer it to less energetic molecules of water. When temperature equalizes over the entire volume, this state obviously corresponds to greater disorder than the ordered concentration of energetic 'hot' molecules in one place and less energetic 'cold' molecules in another place. For this reason processes in nature always tend to level off the temperature. We have already pointed out that this corresponds to the transition to the state of maximum disorder.

The same can be said about the experiment with ink. The probability is much higher in random interactions for molecules to spread over the entire vessel than to gather into a droplet. The uniform distribution of molecules in the vessel with water corresponds to the maximum chaos.

If a process begins from any even partially ordered state, then in the absence of external influences it evolves towards greater disorder.

If we wish to create greater order in a system, we need to exert an external influence on it. For example, we can make heat flow from a cold body to a warmer one. This is what happens in refrigerators: they pump heat from the refrigeration chamber at low temperature to the ambient air, whose temperature is higher than that of the chamber. This, however, requires the work of an electric motor and expenditure of energy.

It is very important that when creating a more ordered state in a system, by influencing it from within a larger system, we inevitably insert additional disorder into this larger system. For instance, the 'pumping' of heat from the refrigerator to the surrounding atmosphere means that the motor produces additional heat in the atmosphere, heating it even more and increasing the degree of 'chaos' of molecular motions in the air. The laws of thermodynamics state that the 'chaos' added to the larger system is inevitably greater than the 'order' introduced into the smaller system. Hence the 'chaos', and entropy, in the whole world must grow, even though order may be established in some parts of the world.

Thomson and Clausius realized that the law they discovered was of exceptional importance for the evolution of the Universe. Indeed, exchange of energy between the world and 'other systems' being impossible, the Universe must be treated as an isolated system. Therefore, all types of energy in the Universe must ultimately convert to heat and the heat spread uniformly through matter, after which all macroscopic motion peters out. Even though the law of conservation of energy is not violated, the energy does not disappear and remains in the form of heat, it 'loses all force', any chance of transformation, any possibility of doing the work of motion. This bleak state became known as the 'thermal death' of the Universe. The reader perhaps agrees that this name characterizes very accurately the very essence of this state.

We know now that the conclusion of the thermal death as inferred by Thomson and Clausius is not applicable to the Universe. The reason is that the Universe is nonstationary, that it has exploded in the past; moreover, an important factor in all processes occurring in the Universe is gravitation. The founders of thermodynamics could not take all this into account. We have

211

already described in detail how the Universe evolves. Violent processes of birth and evolution of worlds take place now as we unfold this story. Nevertheless, we need to emphasize that the conclusion about constant growth of entropy in the Universe remains correct.

The irreversible process in the Universe is thus the growth of entropy. Can this process dictate the direction of flow of time? There can be no doubt that the direction of irreversible processes, as the general tendency for the entire Universe, has something in common with the direction of flow of the river of time. Let us recall, however, that time 'runs' in any process, even the most elementary ones. Note that in all such processes, occurring far away from one another, time ticks on – as far as we can see – in complete accord in the same direction. What causes this accord? Is time affected by the general growth of entropy in the entire Universe? It may be, but so far we know nothing about this influence.

Once you start looking for a global natural phenomenon which might impose the direction of time flow, the expansion of the Universe appears to be the most likely candidate.

Is it possible that the direction of time coincides with the direction of the process of the growth of distances between galaxies in the course of expansion? This idea was suggested by the British theorist Fred Hoyle.

Arthur Eddington even invented a special phrase to indicate the direction of time flow: the 'arrow of time'. Eddington, Hoyle and some others believed that the 'arrow of time' exists, since the Universe does expand. If in the future the expansion is replaced by contraction, then, as these scientists believed, the direction of the 'arrow of time' will be correspondingly reversed.

This hypothesis would deserve discussion if the expansion of the Universe and the recession of galaxies affected phenomena at each point of space. For example, if they resulted in the universal stretching of all bodies and all lengths, if the sizes of stars and planets, of

212

our bodies and atoms and their nuclei grew with the expansion of the Universe. However, nothing of the sort is observed.

The outward motion of galaxies in today's Universe does not affect processes in stars, or their sizes, or the sizes of other heavenly bodies or of atoms of matter. As there is no physical influence, it is difficult to accept that the recession of galaxies may affect the rate of time flow, say, in processes taking place on planets or in reactions of elementary particles. Yakov Zeldovich was decidedly against deriving the direction of the 'arrow of time' from the expansion of the Universe.

The following example is given in the monograph *The Structure and Evolution of the Universe*, written by myself and Yakov Zeldovich:

> Imagine a rocket launched at a velocity lower than the escape velocity... This rocket first moves upwards, away from the Earth, reaches the maximum height and begins to fall down.
>
> Clearly, no laws undergo drastic changes in the rocket: the clock placed inside the rocket keeps ticking monotonously, etc. The transition from expansion to compression in a closed Universe is quite similar to the transition from the ascent of the rocket to descent. It is quite clear, therefore, that the 'arrow of time' does not undergo reversal at the moment of maximum expansion of the Universe. For example, if it did reverse its direction, light rays in the contracting Universe would pour into stars instead of being emitted and lost to cosmic space. Other similarly meaningless examples can be given... in fact, radiation density in the Universe would remain low for a long time after expansion is replaced by contraction, stars would keep emitting light and all local processes in the Universe would continue in the same direction.
>
> The relationship between the 'arrow of time' and expansion is, no doubt, a very important property of our

Universe at the present moment but we cannot use this
relationship to determine the direction of the 'arrow of
time' in the future.

This reasoning is very elementary. The only excuse for
reproducing them here is the insistent repetition of
erroneous point of view in the literature.

Nature demonstrates another sort of process that definitely
'feels' that time flows in one direction only. These are the psychological processes which allow us to sense that time flows from the
past to the future. The direction of this 'psychological arrow of
time' stems from the fact that we remember the past but not the
future.

We have thus looked at three types of natural phenomena that
are patently non-symmetric in time and evolve in a single direction,
at least in today's Universe.

The first class is the class of thermodynamic processes. They
evolve so as to increase chaos and entropy. Such processes define
the 'thermodynamic arrow of time'.

The second phenomenon is the expansion of our Universe: it
gives the 'cosmological arrow of time'.

The third class of phenomena includes our psychological processes that give a subjective feeling of the flow of time. Our memory
of the past and ignorance of the future provide the 'psychological
arrow of time'.

The puzzling thing is the fact that all three 'arrows' point in the
same direction in our Universe of today.

Stephen Hawking discussed this problem in his famous book *A
Brief History of Time* published in 1988. I will reproduce here some
of his arguments, slightly adapting them to the context of this book.

Let us start with the 'thermodynamic arrow of time'. We already
know that this arrow always points in the direction of increasing
disorder, because the number of paths to increasing chaos is always

incomparably greater than that of paths leading to ordering. We can show that the 'psychological arrow of time' must coincide with the 'thermodynamic arrow of time'.

Let us look at how our brain, or its simplified computer model, stores information. A computer is incomparably simpler than the human brain, so let us turn to the operation of its memory. The memory bank consists of a large number of elements which can be in two states. Imagine a device like an abacus with horizontal wires and a bead sliding on each wire. A bead can occupy only one of two positions: shifted either leftmost, or rightmost. It is well known that any message, any information can be written as a sequence of zeros and ones. Let us assume that a bead shifted leftward represents a zero and a bead shifted rightward represents a one. Now it is clear that any information can be written on our sufficiently long 'abacus' by shifting the beads on the sequence of wires to the required positions (right or left). Data can be stored in this form, so this is a 'memory device'. Of course, real computers and our brain are 'technologically' quite different from this memory but the basic principle is the same, and this is all we need to know now.

In order to fix some information in 'memory', we need to shift the beads on the wires in the right way and make sure their positions are correct. All this requires an expenditure of energy which is dissipated as heat (in an electronic computer, it is released both when the computer works and when it cools down). The heat released as a consequence of 'memorizing' makes the ambient air warmer and thus increases 'chaos' (entropy) of the Universe. It is always larger than the order introduced into the storing device when information is recorded. Hawking gives the following example. If you learned by heart each word in a book like this, your memory would record about two million bits of information. This is the measure of how much order was created in your brain. However, reading the book, you have transformed at least a thousand calories of ordered energy stored in food into disordered heat dissipated into the atmosphere. This increases the chaos in the Universe by about twenty million million million million units of data. This is ten million million million times greater than the gain in order in our brain, and that only if you do remember everything in this book...

This means, therefore, that recording information can only increase chaos in the Universe, even though order has been generated in a small corner of the Universe (in the memory storage, on the abacus, in a computer or in the human brain). The sequence in the process in which information is memorized coincides with the sequence in which chaos grows in the Universe. S. Hawking writes:

> The direction of time in which a computer remembers the past is the same as that in which disorder increases.
> Our subjective sense of the direction of time, the psychological arrow of time, is therefore determined within our brain by the thermodynamic arrow of time. Just as a computer, we must remember things in the order in which entropy increases. This makes the second law of thermodynamics almost trivial. Disorder increases with time because we measure time in the direction in which disorder increases. You can't have a safer bet than that!
>
> *A Brief History of Time* (Bantam Books)

To make the argument even more persuasive, Hawking draws the following fantastic picture.

> Suppose, however, that God decided that the universe should finish up in a state of high order but that it didn't matter what state it started in. At early times the universe would probably be in a disordered state. This would mean that disorder would *decrease* with time. You would see broken cups gathering themselves together and jumping

back onto the table. However, any human beings who were observing the cups would be living in a universe in which disorder decreased with time. I shall argue that such beings would have a psychological arrow of time that was backward. That is, they would remember events in the future, and not remember events in the past. When the cup was broken, they would remember it being on the table, but when it was on the table, they would not remember it being on the floor.

ibidem

The White Queen in Lewis Carroll's *Through the Looking Glass* says to Alice: 'It's a poor sort of memory that only works backward'. As follows from the argument above, a 'better' sort of memory simply cannot exist.

The thermodynamic and psychological arrows of time thus must coincide.

However, why should the 'thermodynamic arrow of time' exist at all? In other words, why was the Universe ordered in the past but will develop greater disorder in the future? If the Universe had been in total chaos from the very beginning, this would be the 'thermal death' state, and the Universe would continue to stay dead forever, being slightly disturbed by random fluctuations. There would be no 'thermodynamic arrow of time' in this ubiquitous chaos.

Our Universe is definitely not in this state. What can we say about the degree of ordering in the Universe at the moment of birth? The singularity state should fully manifest the quantum properties of matter and spacetime. This state is therefore completely determined by the quantum properties. What then was the quantum state of our Universe at the moment of birth?

A number of experts, representing various points of view, hypothesized that this state must be ordered to the maximum possible degree. This was suggested by Ya. Zeldovich and L. Grishchuk, a similar hypothesis was formulated by S. Hawking and arguments in favor of this possibility were given in a joint paper by D. Kompaneets, V. Lukash, and myself.

Hawking's approach to this problem is very interesting and original. In order to describe the exceptional state of the Universe at the very beginning of its life, when quantum effects are important for the gravitational field itself, it is very convenient to rewrite the formulas of the theory in terms not of ordinary time but of so-called 'imaginary time'. Imaginary time is obtained by multiplying time by the square root of minus unity. In equations written with imaginary time, this time enters exactly as any spatial coordinate. The time direction in the spacetime now has the same properties as any other spatial direction. Let us use our imagination and draw the directions of imaginary time in the four-dimensional spacetime close to a singularity, that is, in the vicinity of the origin of our Universe. These directions appear as meridians on the globe, converging on the South Pole. The spatial directions are shown as arcs of parallels. Actually, parallels on the globe are one-dimensional while the spatial directions in the Universe are three-dimensional. However, this difference is rather unimportant for us now as far as a visually clear illustration is concerned.

If the Universe's space is closed and begins to expand from a singularity, our illustrative picture (figure 13.1) shows it in the following way. The singularity corresponds to the South Pole. The lengths of the circular parallels demonstrate the size of the closed Universe. The distance along the meridians from the South Pole represents the imaginary time since the beginning of the expansion. The Universe began with zero size at the South Pole (the singularity) and then kept growing – the length of the parallels increased away from the South Pole, reached the equator (this corresponds to the maximum expansion) and this was followed by the contraction phase.

Fig. 13.1.

At this juncture we concentrate our attention on the region close to the South Pole.

In general, there may be a singular point at this point on the surface, such as a sharp peak, but the surface may be quite smooth. Stephen Hawking has suggested that our Universe is the case of just this smoothness. In other words, he hypothesized that the South Pole – the singularity of our Universe, drawn using the imaginary time – is not different in any way from the neighboring points.

The initial state of the Universe must then be of maximum smoothness, that is, it must be ordered. Even though the spatial dimensions of the Universe (the circumference of the parallels) are zero, this point is nevertheless not more singular than, say, the South Pole of our globe. We can in our imagination 'pass' through this singularity on the diagram with imaginary time as through any other point, just as we could travel across the South Pole of the Earth without feeling anything unusual.

Let us now look at the following feature. If we are to the side of the South Pole on our diagram, we easily recognize in which direction the pole (singularity) lies. This is the 'southward' direction, the direction towards the past, towards the beginning of the expansion of the Universe. The future lies 'northward': the further increase of the size of the Universe. Let us now step onto the very South Pole (the singularity). This point has nothing special in any way, except that meridians do start from it. We cannot move southward (to the past) from the South Pole: all paths lead only northward (to the future).

The question of what was before the singularity on this diagram becomes meaningless. Indeed, it contains no 'before' at this point. This is the same as asking what lies to the south of the South Pole: a meaningless question. This example demonstrates a situation in which time is finite, there is no infinitely remote past, but time has no 'beginning', no 'edge'.

Let us turn again to the matter of direction of the arrow of time in our Universe far from the singularity, when the effects of quantum gravity are not important. We should consider our real time rather than 'imaginary time'.

According to Hawking's hypothesis and to the hypotheses of some other theorists, the initial singularity must be smooth. However, this initial state cannot be completely ordered, since otherwise it would contradict the uncertainty relation of quantum mechanics (we touched on this aspect in the chapter 'Journey to unusual depths'). Therefore, there must be at least small deviations from ideal order, some small fluctuations due to the uncertainty relation. This nonuniformity is small at the initial stages of the evolution of the Universe but billions of years later it gives rise to galaxies and creates the large-scale structure of the Universe.

An almost perfect order grows into a greater and greater disorder, which imposes the 'thermodynamic arrow of time'.

Once intelligent beings come on the scene billions of years later, the 'psychological arrow of time' coincides, as we know, with the 'thermodynamic arrow of time'.

How about the third arrow of time, the cosmological arrow which is imposed by the direction of expansion of the Universe, by the increase of its dimensions?

In this era the cosmological arrow points in the same direction as the other two. It is possible, however, that this situation may not always remain unchanged. If the density of matter in the Universe exceeds the critical value, a moment will come in the future when expansion switches to contraction. The cosmological arrow of time will then reverse its direction while the other two will still point in the same direction. The three arrows then cease to be in agreement.

At the beginning Hawking assumed that when the 'cosmological arrow of time' is reversed, the other arrows will also reverse, keeping the trio in agreement. However, he had ultimately to change his opinion and recognize that neither the 'thermodynamic' nor the 'psychological' arrows of time will change their directions.

Hawking asks: 'What should you do when you found you have made a mistake like that?' He gives a clear answer:

> Some people never admit that they are wrong and continue
> to find new, and often mutually inconsistent, arguments to
> support their case – as Eddington did in opposing black
> hole theory. Others claim to have never really supported the
> incorrect view in the first place or, if they did, it was only to
> show that it was inconsistent. It seems to me much better
> and less confusing if you admit in print that you were
> wrong. A good example of this was Einstein, who called the
> cosmological constant, which he introduced when he was
> trying to make a static model of the universe, the biggest
> mistake of his life.

Another example can be added. When Einstein understood that his objections to Friedmann's theory were erroneous, he immediately published a paper in which he admitted his mistake, stated that Friedmann was right, and that Friedmann's work opened new vistas in science.

Finding an error has another side, too. Once you or somebody else have sorted out the physical processes and identified the weak spot, this is also a creative process that elucidates the facets of the phenomenon that were previously unknown, at least to you, or were unclear. A true scientist welcomes the result and does not give way to irritation (even though feelings are never absolutely unambiguous or 'pure', and satisfaction with new understanding may be intertwined with dissatisfaction with oneself, or with other feelings, but a response usually carries a dominant theme). The Moscow physicist Vitaly Ginzburg described how he was analyzing the processes in the so-called transition radiation and came to the conclusion that he had discovered the possibility of creating a very special particle counter. He soon realized his mistake; this is what he wrote: 'The error proved to be of considerable interest, and clarifying it gave pleasure of almost the same degree (albeit very modest) as the 'invention' of the counter itself.'

I should also mention that Andrei Sakharov for some reason favored the idea of reversal of the arrow of time in the Universe. The essential difference between his idea and Hawking's erroneous assumption was that Sakharov believed that the arrow of time 'reverses' at the 'zero moment' of the expansion, not at the point of its maximum expansion (as Hawking did). This means that time flowed backwards before the Universe started the expansion, and that in this backward-ticking time the Universe was also expanding! Sakharov first came up with this idea in 1966, and returned to it later again and again. In all honesty, I could not understand why Sakharov was so fond of this scenario.

222

Let us return to the direction of the arrows of time. The question is: assuming that there may be an epoch in the future when the arrows of time become mis-aligned, why is it that our existence coincides with the epoch in which all the arrows point in the same direction?

The answer may be tied to the so-called anthropic principle. The point is that intelligent life in our Universe could not appear at an arbitrary phase of the evolution of the Universe. It could not arise in the very distant past when there were no stars nor planets and temperature was extremely high. It seems that neither can familiar life forms appear in the distant future when stars burn out or all matter decays. The Universe at the contraction stage is unlikely to resemble today's Universe. If any forms of intelligent life are possible at that stage (I believe that they are), they will evolve completely beyond our recognition. Note that our civilization is very young and life in the form known to us could evolve only on a planet heated by a Sun-like star. We also need to take into account that such stars and planets are only possible at the expansion stage of the Universe, when the matter from which stars are formed still has a store of nuclear energy. The answer is now quite clear.

Being a young civilization, we can exist only at the expansion stage of a universe, with all three arrows of time coinciding in their directions.

Let us summarize.

The direction of time in the world around us is connected with the growth of entropy, growth of disorder, or, to use a metaphor, with aging and decay. If we witness an increase in order in a system, we realize that 'miracles do not happen' (the experience accumulated by science taught us well!) and that this observation indicates that the system interacts with other bodies and that disorder inevitably increases in the totality of these objects. As Ovid wrote in *Metamorphoses*: 'Time, devourer of all things'. Our Universe evolves from the past to the future because it was born in a highly ordered state: 'Past and future are so markedly different because the universe is still very young' (F. Hund, in *Time as Physical Concept*, ed. J. T. Fraser *et al.* 1972, pp. 39–52).

But how about individual 'elementary' processes, such as the motion of a 'pointlike mass' according to Newton's laws? Is the direction of time not specified in them? It appears that the answer to this is that the past and the future do not allow clear distinction in such situations. It is a very different matter how accurately such ideal conditions can be realized, or are realized, in nature.

If we decide to turn to elementary particles of matter hoping to play out 'truly elementary' processes, we encounter a truly puzzling situation.

Elementary particles obey the laws of quantum mechanics, which are very different from the laws of Newton's mechanics. We spent only a little time on these laws in earlier chapters. Nor is this our topic in this chapter. Only a few remarks, important for the topics of time and direction of its flow, will be made.

Quantum theory possesses a profound and well-developed mathematical apparatus. The extraordinary predictions of this theory are shown to hold with unprecedented accuracy; the theory is used by engineers and forms a part of modern technologies. However, even physicists were unable to formulate a universally accepted interpretation of this mathematical apparatus, of the specific images and processes it describes, or even whether one can indeed speak of visualizable images in this field.

I will now try to draw a few strokes outlining the 'image' that, to my understanding, seems to be more plausible. I recommend to readers with more profound interest in these aspects Roger Penrose's book *The Emperor's New Mind* (Oxford University Press, 1989).

To be more specific, let us consider just one example: an electron flying through two vertical slits cut through a screen that is opaque for electrons. If an electron were merely a very small ball obeying the laws of Newtonian mechanics, it would go either through one or through the other hole but would never go through both. Quantum mechanics, whose laws govern the behaviour of the electron, states that its motion is quite dissimilar to the flight of a small spherical object but is a baffling 'mixture' of the two possibilities of passing through the two slits. In fact an electron is, before we try to use some sort of instrument to identify the hole through which it has flown, in some sort of 'mixture' of what went through the first slit and of what went through the second slit. But this is not a simple mixture of, say, 50% going through one slit and 50% going through the other; the proportion is given by a complex number (those who do not know what complex numbers are or dislike them, need not worry, I will not mention them again). This is a very strange image. Can we somehow 'visualize' this strange mixture of states of the electron using proper instruments? No, we cannot. The moment we attempt to determine through which slit the electron moves (anticipating catching it going 'partially' through the first and 'partially' through the second), 'something' happens to the electron and it goes in one jump and *totally* through either the first or the second slit. Physicists speak of a 'reduction' of its state. Strange? Yes, very strange, but true nevertheless.

Does the fact that the electron was in a 'mixture' of states prior to the measurement manifest itself in any way? Yes, it does. If we place behind the screen a plate which records where the electron arrives, and then shoot electrons at it at sufficiently long intervals, the plate will ultimately, after a long exposure, present the famous interference pattern formed by the totality of points where electrons hit, as if a wave was incident on the slits, not individual particles. I will not go into more subtle details of the interference pattern, just say that this pattern is the consequence of the 'mixture of states' of the electron before the measurement.

Is this difficult to comprehend? Yes, very difficult. I believe that hardly any experts will dare say that every detail is now clear. Moreover, different experts suggest different interpretations of observations (and this state of affairs has lasted for almost three-fourths of a century!).

What leaves an especially profound impression is the 'nonlocal' nature of the state of the electron before the measurement. One cannot say (prior to the measurement!) that at a specific moment of time the electron occupies a specific point in space. This is not merely our ignorance: of course, not having performed a measurement, we cannot know where the electron is now. This is more profound: before the measurement the electron *did not* occupy any specific place in space, being a strange mixture of states. I will not go to great lengths proving here why we are so sure of it, since this would lead us too far astray.

Non-locality manifests itself even more strikingly when we look at a system of two or more particles. For example, let two particles first interact and then fly apart to a large distance. If we now perform a measurement on one of these particles, this operation affects the state of the other particle, and this happens, as far as we can see, instantly! Since the effect is not delayed at all, we are not allowed to assume that when the first particle undergoes the measurement, it sends a signal to the second, say, at the speed of light. Not at all: the change in the state of the first particle influences the other particle truly instantaneously! The change is random (stochastic). This is not, therefore, a method of transmitting signals (data) to large distances. The most striking fact is that all these effects have indeed been observed in laboratories in high-precision experiments.

However, let us return to a single electron. As long as no measurements are carried out, its state changes with time according to the well-known laws of quantum mechanics; and these are as reversible in time as the laws of Newtonian mechanics. No direction of time arrow is favored at this stage. At the moment of measurement, however, the state is reduced (a) randomly and (b) irreversibly in time (which is the most important for us now). The two directions of time are thus not equivalent in quantum mechanics.

Some physicists believe, however, that the snag is that the measurements are performed by 'large' instruments, consisting of a huge number of atoms and molecules. These physicists maintain that 'in all likelihood, the root is the "macroscopic nature" of our instruments. If we deal with a tremendously high number of particles of which instruments are composed, the laws of statistics become dominant, just as happens in the case of the law of increasing entropy, which is a corollary of a chaotic interaction of a large number of particles.' This argument may be true but the most important factor appears to be the irreversibility of the reduction process. This process singles out the arrow of time. How and why does this process unfold? We do not know. Experts differ widely in their opinions, so I will not go into speculations. There even exists an opinion that the notion of the 'smooth', continuous (even though curved) four-dimensional spacetime, which serves our needs so faithfully in the study of 'macroscopic' processes, is not adequate for the description of quantum processes.

To illustrate the differences of opinion on the mysterious process of state reduction at the moment of measurement, I will quote two leading physicists, Stephen Hawking and Roger Penrose, who write in their book *The Nature of Space and Time* (Princeton University Press, 1996):

Penrose: 'I call this decay into one *or* the other alternative objective reduction.'
Hawking: 'I totally reject the idea that there is some physical process that corresponds to the reduction... That sounds like magic to me, not science.'

The arguments outlined in this chapter are to a great extent a 'mixture' of reliably established facts, hypotheses that require careful checking, and very fuzzy guesswork.

I want to repeat that we are only beginning to lift the pall of mystery over the nature of time and its mind-boggling properties.

14

Against the flow

Albert Einstein created general relativity theory using a minimum number of experimental data on gravitation; he selected this set of data with the intuition of a genius. Over the many decades since the creation of the theory, all its predictions that allowed observational or experimental verification were invariably proved correct.

Tiny corrections to the motion of the planets of the Solar System, predicted by the theory, were detected and then carefully measured. In 1919 Arthur Eddington discovered the bending of light rays in the gravitational field of the Sun, in agreement with Einstein's prediction.

Then the reddening of light emerging from higher gravitational fields was discovered, which again confirmed Einstein's prediction.

Finally, black holes, those exotic objects that are like nothing else in nature, were discovered – with a high degree of certainty – in the 1970s. In this case, relativity theory manifests itself not in some small corrections to well-known processes but in full-blown effects that drastically change the geometry of space and the properties of time.

Not a single fact that would throw a shadow of doubt on relativity theory was found in all these years. Taken together, the entire experience of science in the 20th century makes one treat seriously the other predictions of the theory, those that have not yet been confirmed by experiment or astrophysical observations.

We have seen that modern physics, which describes the most profound structure of matter, evolves in the direction outlined by Albert Einstein. It is found that all physical processes have a common (unified) nature. It is quite likely that the properties of physical matter are based on complex features of spacetime.

In this chapter we will be talking about the new possibilities predicted by the theory. These possibilities are more than fantastic. But let us face it: modern science reduces the path from science fiction to reality to almost zero length! The reader may dismiss this with a skeptical shrug: 'Whatever you are going to paint is not going beyond rows of formulas written by theoreticians on sheets of paper. This is light years away from any practical consequences.'

The reader is certainly quite right here. There definitely is a very long way from theory to reality. Ludwig Boltzmann once remarked: 'One of my friends defines a practitioner as someone who has no knowledge of theory, and a theoretician as a dreamer who understands nothing at all.'

Nevertheless, I would like to remind the reader that such exotic or extravagant discoveries as nuclear energy or the possibility of space flight became everyday practice in the 20th century. We have already touched on the confirmation of the predictions of relativity theory. Let us remember the wise adage that nothing is more practical than a good theory.

This is the reason why I will talk now about the most daring dreams of physicists, about their most challenging ideas. The famous British physicist J.J. Thomson, who discovered the electron, said that among all the services that can be rendered to science, the most important is the injection of novel ideas.

Actually, I am going to discuss the possibility of traveling to the past.

In summer 1988 Kip Thorne sent me a paper that he and his young students M. Morris and U. Yurtsever had submitted to *Physical Review Letters*. The paper presented arguments in favor of the feasibility – in principle – of moving from the future to the past. This was a very bold paper.

The notion that traveling into the past has been forbidden is embedded in scientific and philosophical thought for a long while now. The reader remembers that travel into the future is a proven fact. We discussed in the chapter 'Time machine' an example of a device that could journey into the very distant future. This was a rocket capable of moving through cosmic space at sufficiently high speed. Returning to the Earth after a flight, the astronaut finds himself in a future epoch of the planet.

There is no doubt that traveling into the future is a very unusual thing for an earthling. For example, if the astronaut spends thirty years in the rocket while the Earth gets older by a hundred and fifty years, he returns younger than his great-grandchildren. Nevertheless, this does not produce any noticeable contradictions. Both the astronaut and the terrestrials were moving in time from the past to the future in habitual circumstances, as always, but the motion was much slower for the astronaut than for the inhabitants of the Earth.

A journey to the past is a very different matter. It seems that if it were possible, we could influence events that happened long ago. This automatically means that we could change the present which depends on the past events that we try to change. Such stories are abundant on the pages of sci-fi publications but until recently physicists were not interested in a serious discussion of the matter.

After the paper was published in *Physical Review Letters* in autumn 1988, the *New York Times* wrote that even if a theoretical demonstration of the possibility of traveling into the past were achieved, this in itself would have profound philosophical and scientific consequences. The traveler in time could, in theory, modify the sequence of events in the past, not excluding his own birth, so that the causality laws on which science is based would be thrown into chaos.

The newspaper then remarked that the authors of the sensational paper reject hypotheses of this sort and state that they work in theoretical physics, not philosophy.

This last remark is perfectly justified: one first has to confirm the theoretical possibility of creating a machine for travel into the past and only after this look into the possible consequences.

Brief notes on the work of the three American physicists flashed in the Soviet media, both in the papers and on TV.

We will now return to the time machine whose prototype has been described in Herbert Wells' short novel with just this title. This was Wells' first science-fiction novel which was published in 1895 and immediately made him famous.

The process of movement through time is interpreted in this novel as something that resembles watching the rapid projection of a movie. The traveler in time, who is not moving and is fixed to the armchair of the time machine, follows the rapidly flicking 'frames' of the consecutive events that move forward in travel from the past to the future or backward in the opposite direction of time travel. Wells gave a superb description of the 'flights' into the future and into the past. Note, by the way, that cinematography was only making its first steps when this short novel was written.

In his youth, Wells showed very serious interest in science, which influenced his early writing as well as all his later work.

The interpretation of the motion through time that the protagonist of the novel offers to his friends impresses me no less than the 'flight' in time. He begins with an obvious remark by the Time Traveller that 'any real body must have extension in *four* directions: it must have Length, Breadth, Thickness, and – Duration' and must exist for at least some interval of time. His conclusion is that this interval is the fourth dimension. He says that 'there are really four dimensions, three that we call the three planes of Space, and a fourth, Time. There is, however, a tendency to draw an unreal distinction between the former three dimensions and the latter, because it happens that our consciousness moves intermittently in one direction along the latter...'

Do not forget that these words were written about ten years before relativity theory was created. Wells' hero then states that different snapshots of the three-dimensional space help study the fourth dimension. 'For instance, here is a portrait of a man at eight years old, another at fifteen, another at seventeen, another at twenty-three, and so on. All these are evidently sections, as it were, Three-dimensional representations of his Four-dimensional being, which is a fixed and unalterable thing.'

We find that history is presented in this description as if in its totality, completely written onto a tape, as in Laplace's interpretation. One can slide along this recording forward or backward. The protagonist of the novel says that the important thing is to learn to slide through time forward or backward just as easily as we do it in space. He points out that we cannot move in all directions in space with equal ease. For example, not so long ago man could not move upward from the surface of the Earth to considerable heights. Furthermore, it is much easier to move downwards, obeying the force of gravity, than upwards. Nevertheless, argues the traveler in time, a person can use a balloon and, defying the pull of gravity, lift himself high above the ground. Why not hope then that ultimately the motion through time may be stopped or accelerated, or even that its direction may be reversed?

There can be no doubt that Wells' novel is a work of fiction, devoted to social problems of the future and giving, to some extent, a warning about how humanity may suffer degradation if it is divided into antagonistic classes. He was, nevertheless, a great artist capable of profound analysis of scientific detail, principles and laws. This is why he achieved an impressive and long remembered description of the dream of flight through time.

Let us return from these dreams to the science of the second part of the 20th century. What can be said from the scientific point of view about the possibility of 'flights' into the past (on the 'flights' into the future, see pp. 70–80).

The first remark is that it is definitely wrong to picture sliding back in time as projecting a movie film in reverse. We will also see that to move through time, it also necessary to move in space (this has already been mentioned on p. 71). Note also that we ourselves cannot get younger in any 'flight' voyage. In any one of us, in any human being and any system, time can only flow forward, only from youth to old age. As Alice says to Humpty Dumpty in Lewis Carroll's *Through the Looking Glass*, 'one can't help growing older'. We know the law of increasing disorder, increasing entropy, which dictates the aging of an organism. (I should make a qualification here that we could fantasize about a purely imaginary situation in which intrusive measures at the live cell level could prevent aging and even bring back youth, but this is a matter of controlling processes in living organisms, not of time flow.) The direction of the 'psychological arrow of time' coincides, as we know, with this 'thermodynamic arrow of time'. Nevertheless, it is possible to imagine that using specially designed machinery, a human being could get into a special 'tunnel' in which he moves backwards with respect to time in the external space, and emerges in the past when passing through the other mouth of this tunnel. Obviously, the traveler through time does not get younger at all. However, having sneaked into the past, he can find himself, for example, in the time of his youth or even in an epoch before he was born!

This journey looks, to a certain degree, like diverting a small fraction of the discharge of a powerful river, pumping this rivulet through a pipe along the bank in the direction opposite to the river flow, and then returning this water to the main flow far upstream.

Something similar to this picture is considered by cosmology today as a theoretical possibility for the river of time. I wish to avoid angering my physicist colleagues who may find out, from one of the readers, what I popularize on these pages; accordingly, I must immediately let everyone know that some of my colleagues are adamant that any travel to the past is definitely forbidden. However, we will return to these disagreements later.

Pure theoreticians, mathematicians rather than physicists, have already dealt for a considerable time with bizarre fantastic worlds in which travel back through time is allowed. These worlds are generated by solving systems of equations of general relativity. It appears that the general opinion has been that these solutions have no connection whatsoever with reality, despite being of great interest for studying the structure of the theory itself. Everyone knows from one's acquaintance with school arithmetic that the formulas of a correct theory can give incorrect – 'physically meaningless' – results. It is sufficient to insert inappropriate numbers into the conditions of a problem, for arithmetic to generate an unacceptable result: say, to excavate a hole in the ground of volume 30 cubic meters in 4 days, with each digger capable of digging 3 cubic meters a day, you need ... 2.5 diggers. Results of this sort made more than one pupil cry. Well, most physicists considered worlds with bizarre properties of time as such 'nonsense' results.

Nevertheless, theorists carefully studied the curious solutions of the equations, even though they recognized their irrelevance. Isn't it indeed curious that theory can conjure up worlds with 'time loops' and where one can sneak into one's own past?

One of the first solutions of this type was obtained by Kurt Gödel in 1949. He considered a stationary, time-invariant universe. For this reason alone, this model could not describe reality, because observations show that galaxies are all flying outward. Gödel's universe is filled with matter possessing rather strange properties, the most important among which is its rotation. The distances between all particles of matter in this universe remain constant in time. If we fixed identical clocks to each particle, we would 'start' them simultaneously for measuring time intervals: for this matter and these clocks the concept of 'simultaneity' is non-existent.

On the whole, the picture of this fantastic universe is quite exotic. This universe is said to contain 'time loops'. From any point in this world one can choose such a path that, moving only forward at a certain velocity, one circles the world and returns to the initial point of the journey exactly at the moment of time at which the journey started! In other words, the traveler circles the world not only in space but also in time.

All this looked very funny. For theorists, this solution was a veritable mathematical toy. No more than a toy, though. It was possible to regard the 'time loops' in Gödel's solution as a funny curiosity, similar to two and a half diggers in the arithmetic problem offered above. In fact, not everybody treated Gödel's result as a light-weight mathematical game. Einstein wrote in 1949:

> Kurt Gödel's essay constitutes, in my opinion, an important contribution to the general theory of relativity, especially to the analysis of the concept of time. The problem involved here disturbed me already at the time of the building of the general theory of relativity, without my having succeeded in clarifying it.
>
> 'Reply to Criticisms', in: *Albert Einstein*, vol. 7 of the Library of Living Philosophers. Edited by P. A. Schilpp (Evanston, IL: Opent Court) pp. 687–688.

My attention was attracted to Gödel's solution by Abram Zelmanov when I was still a student. He himself used this solution as a mathematical example useful in proving a serious theorem. As for me, I simply enjoyed good clean fun analyzing the bizarre properties of curves in this universe.

Theorists also enjoyed 'playing' with other models containing 'time loops'. Zeldovich and myself analyzed one of them in our utterly serious (more than 700 pages long) monograph *Structure and Evolution of the Universe*. This model is very instructive and deserves being briefly outlined here; I hope that it will help to make clearer for the reader what is meant by 'time loops'.

We have already looked in this book at diagrams of spacetime:

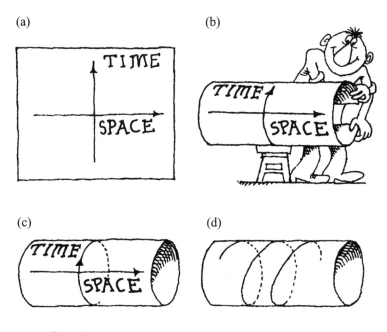

Fig. 14.1.

spatial direction along the horizontal axis, time along the vertical axis. Let us do so again now. Take a sheet of paper (with figure 14.1(a) on it), bend it as shown in figure 14.1(b) and glue the top
of the figure to its bottom. This gives us a cylinder (figure 14.1(c))
on which the circles that form it are 'time loops'. Sliding with the
time flow along a circle on the surface of the cylinder, we return
to the initial moment in the past. It is not necessary, in fact, to
return exactly to the initial event. It is sufficient to move rightward
in space as time ticks on. The lifepath of such a traveler will be
shown by a helix (figure 14.1(d)), it can get longer and longer 'in
time'; in the case of evolution on a closed 'time loop' shown in the
previous figure it had finite length.

In our book, published in 1975, we discussed the fate of physical processes in a universe closed in time in this way. Our emphasis was that in spite of the unusual and rather 'strange' features of processes in this model of the world, it is nevertheless possible to construct a physics in which this situation may not lead to contradictions.

These were merely brief remarks on 'time loops' but Zeldovich and myself treated them in a very different manner. I believed quite seriously that the possibilities of the creation of 'time loops' in a real universe deserved studying. Zeldovich disliked this from the very beginning. Our book was later translated into English and published in the USA. I have recently carefully reread the relevant pages in the English version and discovered, to my dismay, that the description of the model with 'time loops' has been omitted from the text! Zeldovich, unfortunately, died some time ago and I cannot ask him what happened in the process of translation. I guess he simply blue-penciled the paragraphs to which he must have felt an aversion.

Fortunately, there is a passage on 'time loops' in another of my books that was published in English.

I should mention, nevertheless, that I failed to pay the problem of 'time loops' the serious attention that it definitely deserved. This is partly explainable by Zeldovich's skepticism and by the enormous influence that his personality had on me. I did think about 'time loops' long before our book was written, sometimes returned to these ideas later, even tried to calculate some effects, but showed no persistence. Only having read a paper of Kip Thorne and his students, I became excited and burned with a desire to go at least a step nearer to the coveted target: to learn to 'fly' into the past; so I started to work hard on it. Now, what was the gist of the suggestion by the American physicists?

I can separate their work into two stages. Stage one was an analysis of the possibility to create a sort of 'tunnel' connecting two mouths and resembling the gorges in the chapter on 'Holes in space and time'. However, this tunnel must be stable so as to allow passage through it. In other words, this part of the work was the proof that a 'tunnel' can be stabilized against 'collapsing' owing to gravitational and inertial forces.

The second stage was to demonstrate how a such 'tunnel' with two mouths can be converted into a Time Machine.

Kip Thorne recalled that the former problem (how to stabilize a 'tunnel') attracted his attraction after he had leafed through Carl Sagan's novel *Contact*. Sagan was a well-known astronomer and a no less well-known writer. He asked Thorne to check some passages of his new science fiction novel, in which Sagan decided to use black holes for instantaneous transportation of his protagonists to faraway stars. He explained his request by the wish to contradict physics as little as possible. Thorne, who browsed through the novel on his way home in the car, was convinced that black holes are unsuitable for interstellar travel: they offer no exit. However, it appeared that it was possible to use 'tunnels' connecting two black holes if both black holes and the tunnels were stabilized and converted into static structures, passable in both directions. Some mathematical calculations made by Thorne showed that to stabilize a tunnel, it must be filled with an unusual matter or a physical field with properties similar to the vacuum-like state (which has been mentioned earlier in the book).

Then he advised Sagan to correct some places in the novel, which Sagan did when checking the proofs.

Thorne did not let go of this idea. Together with his student M. Morris, he started to work; the first draft of the paper devoted to a (hypothetical) use of stabilized 'tunnels' for fast interstellar travel was ready at the beginning of 1987.

Later, together with U. Yurtsever, they described a more specific device for the 'tunnel'. Let us see what is necessary for gravitational forces not to cause the collapse of the 'tunnel'.

These authors suggest the following system. Begin by creating huge gravitational fields in two not very distant regions of space by compressing masses; this causes strong curvature in these regions (see figure 14.2(a)). Then these regions connect and form a 'tunnel' (another name, invented, as far as I remember, by John Wheeler, is 'wormhole') (figure 14.2(b)). The tunnel created in this way and connecting two mouths is similar to that shown in figure 7.2(a). The difference is in the suggestion of the American scientists to stabilize the tunnel at the moment of its creation. As mentioned above, this is achieved by filling the tunnel with a matter resembling the vacuum-like state (see, e.g. at the beginning of the chapter *Sources*). The antigravitation of this matter prevents the collapse of the tunnel.

Obviously, physicists do not have such matter at their disposal yet, and we do not even know whether the preparation of such matter in the future, in view of its set of properties, can become possible at all. On the other hand, we do not know of any physical laws that prohibit any highly developed future civilization producing such matter. At the moment, the 'details' of how to construct this matter are unclear but special 'bans' on its creation are absent too.

Another, no less fantastic, possibility was outlined by M. Morris, K. Thorne and U. Yurtsever in a paper published in 1988. As you have read earlier in the book, the vacuum on a very small scale is a 'boiling quantum foam', in which wormholes, among other things, are constantly generated and destroyed. The authors wrote: 'One can imagine an advanced civilization pulling such a wormhole out of the quantum foam and enlarging it to classical size.'

Assume, therefore, that this fantastic tunnel, known as a 'wormhole', will some day be constructed. Then one will be able to per-

(a)

(b)

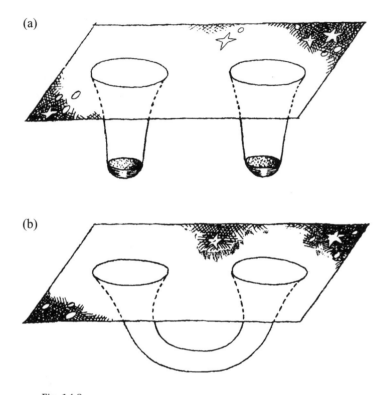

Fig. 14.2.

form the following operation on the mouths that the wormhole connects. They can be pulled away from each other to a great distance without changing the length of the 'wormhole' between them. At first glance, this is quite impossible. It does seem to be so, but only at first glance.

To make this clearer, imagine instead of that space a flat sheet without any mouths or 'tunnels'. Imagine also planar beings that can move in this two-dimensional space among flat stars and conduct geometric measurements. If we now smoothly bend this sheet, as shown in figure 14.3(a), without folding or tearing it, nothing will change on the sheet. All geometrical relations will stay unchanged; the distances between any two points, measured along the shortest lines within the sheet will not change either. It is said that the internal geometry of the sheet, will be conserved. This being so, the planar beings cannot know whether the sheet is bent in any outer space or not. In both cases all 'scenes' on the sheet will be identical. Now imagine that two holes (mouths) on the sheet are connected by a short 'wormhole' (figure 14.3(b)). We now see that a path from one mouth to another in the outer space can be long while that through a 'tunnel' can be short.

This is not all, though. If we pull the upper edge of the sheet, keeping in place the lower edge and both mouths, the upper part of the sheet will slide relative to mouth B in figure 14.3(c). The motion being relative, we can assume that this upper mouth moves among the stars. The distance between the mouths in the outer space can thus change – grow or decrease – while the length of the 'wormhole' will stay unchanged.

All I said about mouths and 'tunnels' on our two-dimensional model surface holds also for mouths and 'tunnels' in three-dimensional space. However, it is infinitely more difficult to imagine this situation in curved three-dimensional space. Mouths A and B seen from outside look very much like black holes. The important difference is that it is possible both to enter and to emerge from them. Seen from inside, they are connected by a tunnel and differ greatly from black holes. The motions from A to B and from B to A are both allowed. The parameters of the mouths and of the 'wormhole' can be chosen in such a way that the gravitational effect on living beings during passage through the 'wormhole' will not be too great and even quite acceptable.

It is now clear how people, having constructed in the future a static 'wormhole' connecting two mouths A and B, could make use

243

(a)

(b)

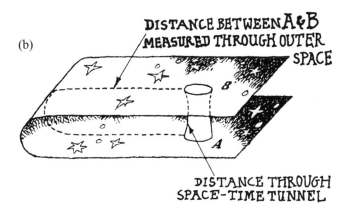

DISTANCE BETWEEN A & B
MEASURED THROUGH OUTER
SPACE

DISTANCE THROUGH
SPACE-TIME TUNNEL

(c)

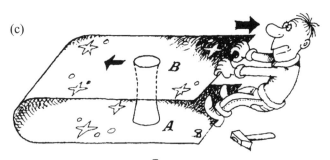

A - STATIONARY
B - MOVING THROUGH STARS

Fig. 14.3.

244

of this device. First we need to tow one of the hole mouths to distant stars without lengthening the tunnel, a condition which will be negligibly short (say, several meters) compared with the distance separating the mouths in outer space (this distance may be thought of as many light years!) (figure 14.3(b)). I need to remind the reader that we have already discussed in the chapter 'Energy extracted from black holes' how to transport black holes through space.

Since the mouths we are talking about as seen from outside are practically indistinguishable from black holes, we can deal with them in the same way.

Now this device can serve as a Space Machine (not as yet a Time Machine). Indeed, a traveler, having entered mouth A in figure 14.3 and having passed through a short tunnel, emerges from mouth B among the faraway stars. The journey may not take much time at all. Reaching the stars will not require a very long and demanding flight through interstellar space.

Even this Space Machine looks splendid and attractive. I hope the reader will forgive me for describing some subtleties of its inner structure; these passages demand certain concentration. The result will compensate for the effort made.

We now switch to the most intriguing part. Let us try and redesign the system of mouths and a 'wormhole' into a Time Machine.

I expect that the argumentation at the beginning of this chapter has made it clear that travel into the past requires a 'time loop'. The mathematical models mentioned earlier all operated with 'time loops' but the models themselves had no relation to reality.

We will now see how a system with a stable short 'wormhole' can produce, according to calculations, 'time loops' in the actual Universe. It could probably be achieved even in the vicinity of our own planet.

The first hints at a project of this type can be found in the Thorne and Morris paper of 1987. Their next paper with U. Yurtsever considerably detailed and improved the project. After reading their paper, I suggested my own version of the Time Machine, also published in autumn 1988. In 1990, together with my colleague and friend Valeri Frolov, I came up with another version of the Time Machine; this is the one I want to present here.

To follow how the Time Machine works, the reader is again invited to show considerable attention and concentration. It can't be helped: we address a Time Machine, no less. Einstein used to say that all things must be simplified as much as possible – but not any further. Those readers who do not care about going into detail are invited to skip a couple of pages and go straight to the result.

Imagine, therefore, two mouths A and B at a considerable distance from each other but connected by a short wormhole. Two identical clocks are placed by the mouths A and B. Owing to the gravitational field near a mouth, both clocks are slowed down relative to clocks far from the mouths. The important point is that this slowdown is identical for both, in view of the symmetry of the picture, but this is unimportant for further discussion and we can safely 'forget' about it for the time being. Note that if we place the clocks somewhat further from each mouth, the slowdown will be quite negligible. The clocks thus tick together.

Now we place our device in a strong gravitational field, say, at the surface of a neutron star, in such a way that mouth B is at the surface while A lies further out, say, at a height of several kilometers.

Now the clocks run differently: the B clock, the closest to the source of gravitation, is slower than the A clock. A very important remark: the relative slowdown is proportional to the distance between the clocks. So far everything seems quite clear: the situation resembles that in the illustration starting the chapter on 'Time, space and gravitation'.

Now we come to the crucial moment. Let us look through the tunnel in the direction, say, from clock A to clock B and try again to compare the readings of the clocks at A and B. The reader may ask 'What for? Haven't we already compared their ticking speeds and found out that B runs slower?' Do not forget, however, that we call Einstein's theory the theory of relativity precisely because it has established the relativity of time. The rate of advance of time depends on the situation, and the one we are considering now is very special. We are looking through a short (several meters long) 'wormhole' that connects spatially very distant places. What do we see? As we know, the slowdown of clock B with respect to clock A is proportional to the distance between them. However, the distance separating them across the wormhole is negligibly small! The clocks thus sit practically side by side. Hence the slowdown of clock B relative to A from the standpoint of an observer or traveler in the wormhole is also infinitesimal.

What does this mean, then? When we look at them from outside, clock B ticks more slowly than A; when we look from inside the wormhole, they tick at identical rates. Which of these two judgments is true then? The reader is certainly ready with an answer: both judgments are true. Indeed, this is a theory-of-relativity situation, and there is no 'absolute rate of time'. Everything depends on circumstances. Both conclusions are thus justified.

We have thus placed our wormhole close to a neutron star and now wait for a sufficiently long time for a sufficiently large difference to accumulate between the readings of clocks A and B (for the external observer). Assume that the difference has grown to two hours (in principle, it could be arbitrarily large).

I stress again: if we look at the clocks through the wormhole (from either of its ends), we always find them showing identical time but if we are in the outer space, clock B is always behind A. Now we tow the two mouths of the wormhole (together with their

247

clocks) away from the neutron star and its gravitational field and 'park' then in an empty spot of the Universe. For convenience's sake, we can now move the mouths close to one another in the outer space, say, at a distance of a hundred meters. We assume that the mouth diameter is, say, several meters. The wormhole length remains the same throughout.

Now, far from the external gravitational field, the clocks are again running at the same rate, but the reading of clock B is behind that of clock A by two hours, because it was closer to the surface of the neutron star and its time then was slowed down. For example, when our observer in the outside space near the wormhole finds that clock A shows 5 o'clock, at the same time he reads off clock B 3 o'clock.

There is nothing super-startling in this yet. I will presently demonstrate, though, that our wormhole can work as a Time Machine.

Let the observer at mouth B glance through the wormhole at clock A. He reads both clocks: B quite close and A farther away, at mouth A. We know that viewed through the wormhole at B, both clocks show the same time as B: that is, 3 o'clock. Now the observer looks at clock A in the outer space (not through the wormhole). He immediately finds that clock A shows 5 o'clock. Hence, when he looks through the wormhole, he observes the past of both clock A and of the surrounding worlds! The observer can walk through the passage and find himself in this past that is younger than the world was around B by two hours (in our example).

This is how this Time Machine works.

To travel into the past deeper than two hours, one has to use a more 'powerful' Time Machine or to pass through the wormhole from B to A two, three or more times. However, this machine only allows the traveler to visit the past in which the Time Machine has already existed. Say, if this Time Machine is indeed designed some day, it will be unable to travel to the Stone Age, since there is no doubt that it did not exist then.

If the observer passes through our wormhole from A to B, he finds himself in the future two hours ahead of the present.

This is what today's physics offers as one version of the Time Machine. One important qualification should be made, though. The geometric size of the Time Machine described above is unlikely to be suitable for actual travel by a human being. The curvature of space (and time!) here is so considerable that huge drops in gravitational forces would tear the fragile human body apart.

For real travel by a human being, the Time Machine must probably be much larger. However, I will not dwell on this, since what is being discussed is the principle, not the details of the design.

Even if the Time Machine is possible, this is work for a highly developed civilization, to which, I hope, mankind will evolve.

There is another point I do not want to evade. From the very beginning of the Time Machine boom, some physicists have risen in arms against it, and there are very well known names among them. Why? As far as I can conclude from observation and discussions, travel to the past would mean a possibility to modify this past, which shoots down the very foundation of science: the causality principle. I do not agree with this, and will discuss it in the next chapter.

For a number of physicists, the first emotional response was later replaced by serious analysis. Preliminary calculations showed that tremendous quantum effects in vacuum could be expected; these would destroy the Time Machine. I do not think that these calculations prove anything since we do not have any consistent theory of these processes and cannot even predict when one might be created. The future will show which of the opponents in this debate were right.

I think, nevertheless, that both sides agree, regardless of disagreements, that an analysis of which processes and events will become feasible and how physics is going to work if the Time Machine is some day created is undoubtedly very important for our understanding of what time is.

15

Can we change the past?

'I'm not kidding you at all, Phil,' Barney insisted. 'I have
produced a workable Time Machine, and I am going to use it
to go back and kill my grandfather.'

'A Gun for Grandfather' by F. M. Busby
in *Getting Home*, (New York: Ace) 1987

I found this epigraph in Paul Nahin's book *Time Machines* (New
York: AIP) published in 1993 and kindly mailed to me. Another
quotation from this book that impressed me with its precision of
analysis is:

Time travel is so dangerous it makes H-bombs perfectly
safe gifts for children and imbeciles. I mean, what's the
worst that can happen with a nuclear weapon? A few million
people die: trivial. With time travel we can destroy the whole
Universe, or so the theory goes.

Millennium
Varley, 1983

Indeed, if a chance to visit the past is available, it seems that by
modifying this past we could modify the lot of some individuals,
the fate of mankind or even the evolution of the entire Universe. Is
this true?

The argument that is especially popular in debates of this sort
is the so-called 'grandfather paradox'. It goes roughly like this: 'If
I could go back into the past in which my grandfather was very
young, I could kill him and thereby make my own birth impossible'.
Or another version of the same paradox: 'I return into my own past,
meet myself in my youth and kill my younger version.'

In both cases this unnatural homicide generates complete nonsense. Should we infer that such an event is impossible? But why?
I have my 'free will', don't I? Hence I can realize this 'free will', at
least in principle.

Science fiction writers have scrutinized all possible versions of
this scenario. Those readers who enjoy literary fantasizing (which

is sometimes quite engrossing) can be referred to the above-mentioned book by Paul J. Nahin, which offers a huge collection of references. But here we return to physics.

Does the 'grandfather paradox', or other similar paradoxes, demonstrate that travel through time is not allowed? Indeed, it seems logical that having gone back in time and eliminated the cause of a phenomenon that has already taken place in the present, we thereby violate the fundamental principle of science: causality!

But is this true? I doubt it, and suspect that the argument as given above is flawed. What has physics to say about the likely consequences of meeting oneself (or one's grandfather) in the past?

Obviously, a physicist (at least our contemporary physicist) is unable to perform an exact calculation of the actions of a human being. This is the field for psychology and sociology, not for physics. However, a physicist can give a strict calculation of what happens to simple physical objects after they pass through a time machine. Let us use such simple objects, model the grandfather paradox and see how it can be resolved.

Before starting on this exciting journey, I wish to attract the reader's attention to one totally new factor that arises here. If a 'time loop' exists, the events on this loop cannot be separated into future and past. To clarify this statement, let us consider the following example.

Imagine that I walk in a long string of people moving along a straight line. I can definitely say which of them is in front of me and who is behind. If, however, we all follow a circle, I can say 'ahead of me' or 'behind me' only about my nearest neighbours but not about the entire line of people. Regarding people further and further ahead of me, I ultimately go around the entire circle and reach myself from behind. This is why people moving on a circle cannot be divided into those 'moving in front' and those 'walking behind'.

The same is true for the 'time loop'. We can say which of the nearest events belong to the future and which to the past. But this division cannot be applied to the time loop as a whole. The loop has no clearly defined future and no past, and all events affect one another on a circle. Briefly and metaphorically speaking, we are under 'double' strong influence: without the time machine events are influenced by the flow of data from the past (but not from the future! this is the gist of the causality principle), while events on the loop respond to information coming from both the past and the future.

Therefore, with the time machine, today's events must be consistent with (i.e. be determined by) not only the past but also the future! I formulated this self-consistency principle many years ago and now it appears to have been accepted by everyone who works in the time machine field. Recently I and my colleagues were able to prove that this principle can be deduced from the fundamental laws of physics.

Let us return to modeling the 'grandfather paradox'. Consider the following simple example: the motion of a billiard ball on a table, assuming no time machine exists.

Pushing a ball, it is not difficult to direct its motion into a chosen pocket A (see figure 15.1(a)). Now take another ball, identical to the first, and send it rolling before the first ball reaches the pocket, in such a way that it collides with the first ball at point C in figure 15.1(b). This is usually not a problem even for a moderate-class billiards player. Now the first ball changes its trajectory after the collision and no longer goes into pocket A.

We can say that the 'fate' of the first ball changed dramatically because of the collision with the second ball. It remained on the table instead of passing through the mouth A and dropping into the pocket.

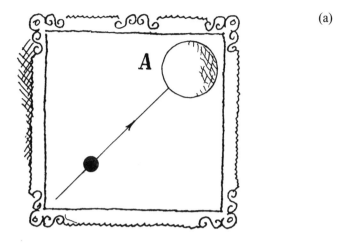

(a)

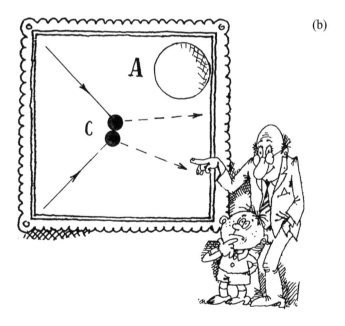

(b)

Fig. 15.1.

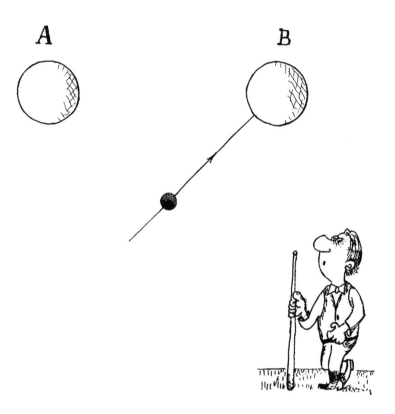

Fig. 15.2.

Now let us run a similar experiment but make use of a time machine. In contrast to the previous experiment, this one involves a single ball. In addition, we will run it not on a billiards table but in cosmic space, far from the gravitational field of the Earth.

Assume that, as shown in figure 15.2, we have a time machine with two mouths A and B. If the ball enters mouth B and then passes through the short wormhole, it emerges from mouth A *in the past*, before it entered mouth B.

Assume now that our time machine is not very 'powerful' and

sends the ball into the past only 20 seconds back. We can begin our experiment. Let us send the ball by the cue from some distance to mouth B. We know that having entered mouth B, it will emerge from A in the past, 20 seconds before it enters B. The picture therefore looks like this.

The ball moves towards B but before it enters the mouth, its 'older version' emerges from A from the future and keeps moving outside the time machine (see figure 15.3(a)). It is not impossible to calculate the force and direction of the original cue push in such a way that both versions of our ball, the younger ball (i.e. the ball before it sank into B) and the older ball (i.e. the ball arriving from the future through mouth A) will arrive simultaneously at a point C and collide.

Now everything resembles what we normally see on billiards tables. As a result of the collision, the younger version of the ball sharply changes its trajectory, so the ball will never enter mouth B of the time machine.

'Nonsense', says the reader. 'If the ball does not fall into B, its 'older version' will never emerge from A! This means that the collision will not occur and the younger ball will roll into B and emerge from A. This will produce a collision!... The paradox is staring us in the face!'

In fact, we made an elementary logical mistake in this chain of arguments. Indeed, when we followed the trajectory of the younger ball towards mouth B, we first chose to ignore the effect of collision. Only after making sure that the collision would happen, we said: 'Now we will take into account this collision and see how the trajectory has changed after the impact.' However, this is a flawed way to argue. The effect of the collision should have been taken into account *from the very beginning*. Indeed, the ball moves once only, and we cannot treat its motion as collisionless once, and then as motion with collision. This means that the effect of the future

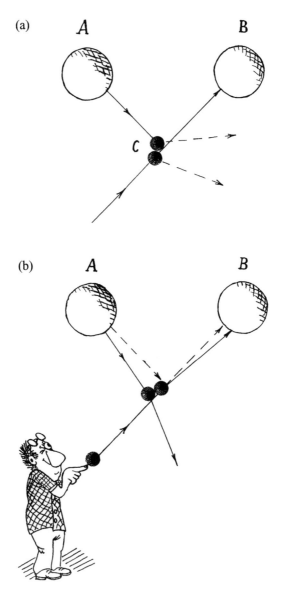

Fig. 15.3.

(i.e. of the older ball arriving from the future) on the event must be considered from time zero.

What is going to happen if we immediately include into consideration the effect (impact) of the ball from the future (the older ball) on the younger ball in the present? Here we go.

Imagine that the collision was not severe, just a slight glancing blow. The reader may be baffled: 'Can't we adjust the momentum and direction of the impact by the younger ball moving towards B in such a way that the collision is violent, not just glancing?' The reader will know what happens in this situation just a few lines later.

First we need to analyze what happens with the glancing collision. The trajectory of the younger ball then does not change drastically, but only a little bit (see figure 15.3(b)). It enters mouth B, emerges from A in the past and then moves along a trajectory which is slightly different from the one on which the motion of the younger ball was ignored. On this modified trajectory the older ball collides with the younger ball not too violently (as it would if the changes were ignored), just a slight glancing blow (see figure 15.3(b)).

Here lies the rub! Taking into account the changes in the trajectory in response to the collision may automatically result in a glancing blow! A straightforward mathematical calculation shows that if we choose the direction and momentum of the initial push in such a way that by our initial assumption the collision is violent and leads to a paradox (i.e. we ignore the consequences of the collision when we plan the initial push), in reality we obtain a weak, glancing collision.

Is it possible to organize a violent collision between the younger and older versions of the ball by properly adjusting the initial conditions? Yes, it is. To do this, we need to push the ball in a direction very different from the direction to mouth B, so that without the time machine the ball would roll far from B. However, with the time machine the older version of the ball emerges from mouth A, hits the younger version a strong blow and changes its trajectory exactly to send it into mouth B and allow it to emerge from mouth A in the past. Note that all this happens quite automatically! Without any deliberate tuning of the initial conditions for the initial push: these conditions are quite arbitrary.

I will not go here into many other exciting features of this problem. The reader may notice that even in the case of the simple mechanical motion of a single (!) billiards ball the picture with a time machine differs dramatically from an ordinary problem without one. The important thing is that the laws of physics *automatically* prevent the paradox.

'Very well', says the reader, 'A paradox may indeed be removed in the example analyzed above. However, let us set a slightly more complicated experiment: let the ball fly through space, not roll on a table, and let us install in the forward-facing part of the ball an automatic radar- and computer-controlled cannon. This gadget triggers the gun and the gun fires when the radar detects a billiards ball at a short distance ahead of the armed ball (say, less than a meter ahead).

Now the scene is very similar to the grandfather paradox. Once the ball emerging from the future comes close to the point C of collision with its younger version, the cannon on the older ball shoots and blasts the younger ball to pieces. No glancing blows are possible now, so the paradox seems to be inevitable.'

Patience, my reader, patience; we will see soon enough that no paradox is produced in this case either.

One of the versions of what could happen looks as follows.

The younger ball with a cannon that points forward moves towards mouth B. The older version with a cannon emerges from A and moves to the rendezvous point C. However, this older version rotates, in contrast to the younger version. (I will explain several sentences later how this rotation arises.) When the two balls arrive at the meeting point, the cannon and radar of the older ball point sideways, owing to rotation, the radar fails to 'see' the younger ball, the device is not triggered and the collision is glancing again. The blow slightly changes the trajectory of the younger ball and sets the ball rotating. This is the reason why the ball emerging from the future at A was rotating. Rotation prevented the catastrophe and eliminated the paradox.

This example gives a fairly accurate simulation of the 'grandfather paradox', even if in a very simplified mechanistic form.

The reader may now counter like this: 'Fine, but I still believe that the paradox is inevitable under certain circumstances. Let us design something really drastic. For example, let us put into the ball a bomb that will explode if any point on the surface of the ball is touched. It seems obvious that an explosion destroying both balls will occur in any collision, however glancing it may be. The paradox is inevitable. What can you say to this?'

The catastrophe is probably unavoidable, but the paradox is not produced all the same. Events develop as follows (see figure 15.4). The younger ball with a bomb moves towards B. At a certain moment mouth A lets out ... oh no, not the older version of our ball, but a fragment ejected in the explosion. It will be clear, again several sentences later, where the fragment came from.

This fragment flies through the outer space and hits the younger ball, triggering the explosion. The ball explodes, fragments fly in all directions, and the fragment that caused the explosion is destroyed as well. At least one fragment enters mouth B, moves through the corridor, and emerges from A in the past. This is the fragment that caused the explosion.

This example demonstrates with special clarity how the future may define the events in the present and how, in the presence of

Fig. 15.4.

the time machine, the future, the past and the present are, in fact, 'mixed together'. Indeed, a fragment, which is a result of the explosion of the bomb, passes through the time machine, enters the past and becomes the cause of the explosion itself!

Let us make a stop here. Numerous problems can be thought up. Some of them produce even more paradoxical results than those we have traced, others cannot be calculated because of their complexity. Nevertheless, no proof is known that even one of these examples leads to contradictions. In my opinion, no such proof exists.

The time machine drastically changes quite a few processes and leads to most unexpected consequences. I will not analyze them here. Physicists have only started the work in this field.

Let us recapitulate.

With the time machine becoming a reality, the future starts to affect the past. All events occur in such a way that this influence is taken into account. However, once an event has taken place (it was influenced by the events both in the past and in the future), that's the end, it cannot be altered. 'What has already happened cannot be undone' (Amelia Greene, 1983).

Still, how about the assassination of the grandparents? Could this extravagant crime be committed using the time machine? The answer is a categorical *no*. Kip Thorne puts it this way:

> ...something has to stay your hand as you try to kill your
> grandmother. What? How? The answer (if there is one) is far
> from obvious, since it entails the free will of human beings.
> The compatibility between free will and physical law is a
> terribly muddy issue even in the absence of time machines.

As for the constraints on 'free will', the reader should notice that even without a time machine, *any law of physics* places limits on 'free will'. Say, I might wish to walk on the ceiling (without special

263

equipment): my 'free will' prompts me to. This, however, is forbidden. The law of universal gravitation limits my 'free will' and there is nothing I can do about it.

In the presence of the time machine the constraints on 'free will' are, of course, somewhat different, but they are not, in principle, anything extraordinary in the physics of our time.

I will conclude this brief discussion of the limitations imposed on 'free will' with a remark made by Einstein and which may be of interest to those readers who find time to mull over problems of this type. 'Schopenhauer once remarked: "A man can do what he wishes but he is not free to wish what he wants".' (*Epilogue. A Socratic dialogue* in M. Planck *Where is Science Going?* London, 1933, p. 210).

It must again be emphasized that some physicists flatly reject any work on matters connected with the time machine.

Only future research will show who was right in this. Shakespeare said:

> And enterprises of great pith and moment
> ... their currents turn awry
> And lose the name of action

I am an optimist and believe in the enormous promise of this new field of research.

As for the practical realization of the new ideas, I would like to conclude this rather complicated chapter by remembering that Wilbur Wright wrote in 1901 that man would learn to fly in not less than a thousand years. Nevertheless, the first flights of Orville and Wilbur Wright were achieved in 1903, and now we have reached the planets!

Conclusion

Despite all this, in spite of all the progress, the nature of time remains to a large extent a mystery for us. Regardless of the millennia counted by the history of science, we are only at the very beginning of the way to comprehend the essential meaning of time flow.

Our knowledge about this 'grand river' was gleaned very slowly. The science of the ancient Greeks defined the concept of time as an independent category, as a universal property inherent in all objects and phenomena of the material world. It also established that time does not move in circles, that it is not cyclic, that it moves inexorably from the past to the future.

The laws of classical physics, which found their exhaustive expression in Newton's work, assigned to time the role of empty duration, without beginning or end, flowing eternally at a constant rate, regardless of what events take place in the world.

The revolution in physics that began a century ago, and the subsequent relentless progress in this science produced, numerous overwhelming discoveries. We now know that the rate of flow of the river of time can indeed be influenced. In principle, a 'flight' can be made to the very distant future, and, who knows, one may be able to move 'upstream' on the 'time river', that is, into the past;

technically, of course, both sorts of time travel remain unfeasible at present.

Science has discovered that the properties of time at the very first moments after the birth of our Universe were very different from what we observe today. Time existed then as individual quanta. Time inside black holes discovered by astronomers in the Universe is also extremely unusual. Time at the very core of a black hole also splits into 'droplets'. Physics is slowly beginning to understand better why time flows continuously and can never stop.

But the deeper the penetration of science and the more mysteries it clarifies, the more problems it discovers that are less predictable and even more daunting. In this book I have attempted to describe only one, extremely exciting direction of this perpetual motion.

I feel greatly surprised that some outstanding physicists, among our predecessors and our contemporaries, have held the opinion that the road filled with productive discoveries of new laws is not endless, that all the most important physical laws will some day, sooner or later, be known. For example, Richard Feynman, one of the creators of quantum electrodynamics, whose contribution brought him the Nobel prize for physics in 1965, wrote:

> ... I think there will certainly not be perpetual novelty, say for a thousand years. This thing cannot keep going so that we are always going to discover more and more laws. If we do, it will become boring that there are so many levels one underneath the other.... We are very lucky to live in an age in which we are still making discoveries. It is like discovery of America—you only discover it once. The age in which we live is the age in which we are discovering the fundamental laws of nature, and that will never come again. It is very exciting, it is marvelous, but this excitement will have to go.
>
> *The Character of Physical Law*
> (London: Cox and Wyman) 1965

I suspect that the driving stimulus for a mood of this sort could be the feeling that a grandiose historical period in physical research is coming to an end; such was the completion of the period of classical physics of Newton and Maxwell, or the completion of the creation of quantum electrodynamics.

However, when any stage in physics ends, even if it was unusually brilliant, a new stage starts. Most physicists disagree with the opinion of the possible end of science. I will quote a well-known Moscow physicist, Moisei Markov, who said that 'what we discern in front of us is a truly new and in a certain sense resplendent era in science'. I firmly believe that an important role during this era will be played by the study of the mysteries of space and time.

The task facing scientists now is to understand why time is uniform, what the relations are between its properties and the general properties of the Universe. Finally, we are approaching tackling the problem of implementing 'flights' through time to both the future and the past.

As long as humankind exists, it will strive for new knowledge and will make new discoveries.

Name index

Subject index